Networking the Globe

Contemporary events that have catastrophic global ramifications, such as the current economic crisis or on-going conflicts across the globe, are not only mediated by super-fast digital communication and information networks, but also conditioned by the presence of rapidly advancing technologies. From social network sites like YouTube and Facebook to global satellite news channels like Al Jazeera or the BBC World Service, digital forms of culture have multiplied in recent years, creating global conduits and connections which shape our lives in many ways.

Bringing together an interdisciplinary group of scholars, this book addresses how new technologies have impacted discussions of identity, place and nation, and how they are shifting the parameters of postcolonial thought. Each chapter reflects on current research in its respective field, and presents new directions on the interconnection between new technologies and the postcolonial in a contemporary context. Offering a major intervention in debates around global networks, this thought-provoking collection highlights innovative research on new technologies, and its impact on a 'postcolonial' world. This book was originally published as a special issue of the *Journal of Postcolonial Writing*.

Florian Stadtler is a Lecturer in Global Literatures at the University of Exeter, UK. He has published on South Asian cinema, fiction and history, including *Fiction, Film and Indian Popular Cinema: Salman Rushdie's Novels and the Cinematic Imagination*. He is the reviews editor for *Wasafiri: The Magazine of International Contemporary Writing*.

Ole Birk Laursen is a Postdoctoral Research Fellow at the University of Copenhagen, Denmark. His research concerns the literature and history of anti-colonial and postcolonial resistances in Britain, focusing especially on anarchism, revolutions and riots.

Networking the Globe

New technologies and the postcolonial

Edited by
**Florian Stadtler and Ole Birk Laursen,
with Brian Rock**

Routledge
Taylor & Francis Group

LONDON AND NEW YORK

First published 2016
by Routledge

2 Park Square, Milton Park, Abingdon, Oxfordshire OX14 4RN
711 Third Avenue, New York, NY 10017

Routledge is an imprint of the Taylor & Francis Group, an informa business

First issued in paperback 2017

British Library Cataloguing in Publication Data
A catalogue record for this book is available from the British Library

ISBN 13: 978-1-138-94589-0 (hbk)
ISBN 13: 978-1-138-30543-4 (pbk)

Typeset in Times New Roman
by diacriTech, Chennai

Publisher's Note
The publisher accepts responsibility for any inconsistencies that may have arisen
during the conversion of this book from journal articles to book chapters, namely
the possible inclusion of journal terminology.

Disclaimer
Every effort has been made to contact copyright holders for their permission to
reprint material in this book. The publishers would be grateful to hear from any
copyright holder who is not here acknowledged and will undertake to rectify any
errors or omissions in future editions of this book.

Contents

CONTENTS

Citation Information

The chapters in this book were originally published in the *Journal of Postcolonial Writing*, volume 49, issue 5 (December 2013). When citing this material, please use the original page numbering for each article, as follows:

CITATION INFORMATION

Chapter 6

Pluralism and cultural imperialism in the network films Babel *and* Lantana
Vivien Silvey
Journal of Postcolonial Writing, volume 49, issue 5 (December 2013) pp. 582–595

Chapter 7

The global and the postcolonial in post-migratory literature
Ahmed Gamal
Journal of Postcolonial Writing, volume 49, issue 5 (December 2013) pp. 596–608

Chapter 8

The cartography of the local in Arun Kolatkar's poetry
Anjali Nerlekar
Journal of Postcolonial Writing, volume 49, issue 5 (December 2013) pp. 609–623

For any permission-related enquiries please visit
http://www.tandfonline.com/page/help/permissions

Notes on Contributors

Ramy Aly is a Lecturer in Anthropology at the University of Sussex, Brighton, UK. His Ph.D. thesis 'Be(com)ing Arab in London: Performativity Between Structures of Subjection' was based on fieldwork done in London between 2006 and 2007 among young people born or raised in London whose parents were migrants from Arab States. He is author of the chapter 'Producing Men, the Nation and Commodities: The Cultural Political Economy of Militarism in Egypt', in *Militarism and International Relations: Political Economy, Security, Theory* (edited by Anna Stavrianakis and Jan Selby, 2012).

Sandra Annett is an Assistant Professor specializing in digital and new media studies in the Department of English and Film Studies at Wilfrid Laurier University, Waterloo, Canada. Along with her university teaching, she researches and delivers occasional lectures in Japan, most recently at Wakō University, Tokyo. Her current research interests include media studies, globalization and postcolonial theory, East Asian popular cultures, animation and fan studies.

Tracey Black has a Ph.D. in Anthropology from University College London, UK, and is currently pursuing research in South Asian visual culture.

Paula Blair is a Teaching Fellow in Film at the University of Newcastle, UK. She is an early career researcher in film and visual culture. She received her Ph.D. from Queen's University Belfast, UK. Her book, *Old Borders, New Technologies: Reframing Film and Visual Culture in Contemporary Northern Ireland* (2014), was developed from her Ph.D. thesis, which won the Peter Lang Young Scholars in Film Studies 2012 Award.

Ahmed Gamal is a Professor of English and Comparative Literature in the Faculty of Arts, University of Dammam, Saudi Arabia. He was a Fulbright Visiting Scholar at Columbia University, New York City, USA, in 2010–11.

David Herbert is Professor of Religion and Society at the University of Agder, Kristiansand, Norway. His main works are *Religion and Civil Society* (2003), *Religion and Social Transformations* (ed., 2002), *Creating Community Cohesion: Religion, Media and Multiculturalism in North West Europe* (2013), and *Social Media and the Sacred* (edited with Marie Gillespie and Anita Greenhill, 2013).

Maruta Herding is a researcher at the Deutsches Jugendinstitut, Munich, Germany. She is a sociologist with a particular interest in cultural sociology, anthropology and qualitative research. In 2012, she completed her Ph.D. at the University of Cambridge, UK, on Islamic Youth Culture in Western Europe.

NOTES ON CONTRIBUTORS

Ole Birk Laursen is a Postdoctoral Research Fellow at the University of Copenhagen, Denmark. His research concerns the literature and history of anti-colonial and postcolonial resistances in Britain, focusing especially on anarchism, revolutions and riots.

Anjali Nerlekar is an Assistant Professor at Rutgers University, New Brunswick, NJ, USA. Her areas of research include South Asian poetry, with a focus on post-independence Indian poetry; translation studies; South Asian literature; Indo-Caribbean literature; postcolonial book history; and world anglophone literature.

Vivien Silvey is a Learning Adviser in the Academic Skills and Learning Centre at the Australian National University, Canberra, Australia. She was recently awarded a Ph.D. from the Australian National University for her thesis on network narratives, genre theory and comparative studies in world cinema.

Florian Stadtler is a Lecturer in Global Literatures at the University of Exeter, UK. He has published on South Asian cinema, fiction and history, including *Fiction, Film and Indian Popular Cinema: Salman Rushdie's Novels and the Cinematic Imagination*. He is the reviews editor for *Wasafiri: The Magazine of International Contemporary Writing*.

Hilde C. Stephansen is a Lecturer in Sociology at the University of Westminster, London, UK, where she teaches across a range of course modules covering introductory sociology, cultural studies, contemporary social theory, research methods, media and globalization. Her current research focuses on global media activism and its implications for how we might understand the concept of the public sphere.

INTRODUCTION

Networking the globe: culture, technologies, globalization

Florian Stadtler[a], Ole Birk Laursen[b] with Brian Rock[c]

[a]University of Exeter, UK; [b]Open University, UK; [c]University of Stirling, UK

Contemporary events have highlighted important connections between technology, globalization and cultural production. Information technologies in particular have impacted the global commodification of information and have led to the significant erosion of national boundaries – for example, through Internet forums and self-broadcasting. Access to these technologies has influenced local and global identities, especially youth cultures through digital platforms as subcultural expression, creating virtual subjectivities and transnational communities. These technologies have also facilitated a wider global network and interconnection of narrative forms and genres that have led to explorations of alternative modernities in a globalizing postcolonial context. This introduction highlights the need to enrich the intellectual resources being brought to bear upon the development of contemporary information technologies through postcolonial theoretical frameworks. By addressing the rise of new media technologies in postcolonial spaces, it focuses on how the use of such modern technologies has enabled transnational networks of cultural, social and political exchange.

Since the new millennium, there has been a succession of crises with catastrophic global ramifications, such as the economic downturn and conflicts across the globe, not only mediated by super-fast digital communication and information networks, but also conditioned by rapidly advancing technologies. From social networking sites such as YouTube and Facebook to global satellite news channels including Al-Jazeera and the BBC World Service, digital forms of culture have multiplied in recent years, proliferating conduits and connections across the globe; these shape our lives in multifarious ways. In the light of this, a postcolonial perspective on information and communication technologies is pressing. How far is cyberspace mediated by metropolitan centres of knowledge production, and how might new media entrench existing structures of inequality, by serving corporate capitalist interests or by saturating consumers with hegemonic representations of culture or global events? Conversely, to what extent can technologies operate as tools of empowerment and resistance for marginalized peoples, by bypassing forms of censorship and facilitating access to global arenas of debate and alternative communities?

The articles in this special issue stem from the inaugural postgraduate conference of the Postcolonial Studies Association held at the University of Stirling in May 2010, which provided an incisive forum for wide-ranging debates on the issue of new technologies and global networks in postcolonial contexts. This discussion, reflected in this issue, was shaped by contributors' work in a wide range of different disciplines, including anthropology, film, screen, media and cultural studies, sociology and literature, exploring in exciting ways the impact of postcolonial concerns in other fields. This meant that a wide array of topics was covered and approached from different theoretical angles and methodologies.

Hilde C. Stephansen's article on the process of the World Social Forum explores activists' use of communication tools to challenge traditional conceptions of centre–periphery divisions. Focusing her case study on a poor urban community in the south of Brazil, she elaborates further on the epistemological significance of place in relation to contexts of globalization.

David Herbert, Tracey Black and Ramy Aly engage with issues of technology and communication, especially in the context of transnational media and its social media facilities which engage their audiences. They compare the content, context and facilitation of discussions about religion and politics on contrasting types of Internet forums linked to the current affairs output of the BBC World Service (BBCWS), and with rival services in English and Arabic (Al-Jazeera and Al-Arabiyya). They question the extent to which these Internet forums linked to the channels' programming output can enable contraflows to counteract centralizing tendencies and, in so doing, highlight the role the media plays in offering space for social and political commentary.

Paula Blair takes up the issue of national boundaries but develops it in the context of cultural practices in visual art. Central are her deliberations on issues of representation in an insightful discussion of surveillance in video art installations in Northern Ireland. She challenges notions of surveillance as knowledge acquisition and a tool of power in the context of the Troubles and the peace process. Drawing on the Foucauldian notion of panopticism, social control, and the ambivalence of surveillance in western society, she argues that "advancing technologies allow for different kinds of invasion/occupation, and technomediated modes of modern living come complete with adapted exclusions and inclusions".

Maruta Herding's article redirects our attention to Islamic youth cultures in Europe. In a comparative analysis of British, French and German digital platforms, she explores the role new media plays in Muslim youths' identity formations. Though recognizing its transnational potential, her evidence suggests that thus far there is little cross-border interaction between participants, or awareness of similar trends in other European countries outside their own. In this respect, their responses are shaped by a personalized experience in national contexts of their home countries in the west.

Sandra Annett explores the complex network of exchanges that take place in animation, and how advances in technology impact relations and collaborations across cultures. In particular, she focuses on how film production facilitates new global interconnections and impacts on patterns of consumption, viewing practices and responses. In so doing, she highlights how new media platforms are reworked for differing audience constituencies in national and diasporic contexts that lead to mutual yet asymmetrical exchanges.

Vivien Silvey explores a different narrative mode through network paradigms in the context of two 21st-century films in world cinema, *Babel* (2006) and *Lantana* (2001), and their representations of cultural pluralism. She argues that ultimately both films are compromised by their narrative politics and thus fail in their aim to give equal voice to marginalized sections of society.

We conclude with two articles which reorientate the issue of globalized networks towards literature.[1] Anjali Nerlekar examines the poetry of Arun Kolatkar. She focuses in particular on his mapping of Bombay and his exploration of the cityscape in the context of economic liberalization in India. This positions the city in a wider global network while also, contradictorily, occupying a localized marginality.

2

Ahmed Gamal engages with two post-9/11 novels, *The Reluctant Fundamentalist* (2007) by Mohsin Hamid and *Burnt Shadows* (2009) by Kamila Shamsie. He explores the transcultural contact zones of these fictions in the contexts of theorizations of "post-migratory" literature to consider wider issues of globalization in relation to the postcolonial.

The contributors to the issue address not only specific texts, contexts and case studies from their respective fields, but consider new theoretical directions and debates in relation to new technologies, global networks and the postcolonial as a whole. From the articles a coherent set of questions emerges, relating both to broader interconnections and to areas of disengagement between new technologies in a globalizing world and their influence in postcolonial locations. In so doing, we hope the issue makes a valuable contribution to ongoing debates around global networks and underlines the impact of technologies in different territories – and its global consequences.

Acknowledgements

We would like to acknowledge the University of Stirling, the Open University and Birkbeck, University of London, for supporting the conference. Thanks are also due to Rehana Ahmed for co-conceptualizing and co-organizing the conference and her input to this special issue. We also gratefully acknowledge the journal editors, Janet Wilson and Chris Ringrose, for their advice, guidance and support in bringing this issue to fruition.

Note

1. Please note that these two articles were accepted for publication before the *Journal of Postcolonial Writing* switched to *Chicago Manual of Style* and therefore the citations and reference lists remain in MLA style.

Connecting the peripheries: networks, place and scale in the World Social Forum process

Hilde C. Stephansen

The Open University, Milton Keynes, United Kingdom

Communication technologies occupy a central place in contemporary theorizations of transnational social movement networks. Not only does the internet provide the technical infrastructure through which activists communicate and share information, increasing their capacity to introduce oppositional messages into the public realm (Castells); its network architecture is also closely linked to the organizational logic of contemporary social justice movements (Juris). While recognizing the fundamental importance of communication technologies for such movements, this article cautions against overly disembodied conceptions of transnational activist networks and highlights the need to pay attention to issues of place and scale, as well as the importance of affect in the construction of alternative global imaginaries. Through a case study of a small social forum event held in February 2010 in a poor urban community in the south of Brazil as part of the World Social Forum process, the article examines activists' use of communication technologies to construct transnational networks between different place-based actors. It shows that these practices are not simply concerned with establishing links between already existing places; the creation of networks is also inextricably bound up with particular constructions of place. By engaging in a politics that is simultaneously place-based and global in scope, these actors challenge traditional conceptions of scale as well as dominant epistemological paradigms.

Introduction: the World Social Forum

The World Social Forum (WSF) is widely recognized as one of the most important manifestations of what might be referred to as the "global Left" (Santos 2006). First organized in 2001 in Porto Alegre, Brazil, the WSF was originally conceived as a counterpoint to the World Economic Forum, which annually gathers the world's political and economic elites in Davos, Switzerland. The WSF is currently held biennially in different locations around the world (thus far, always in the global South) and regularly brings together tens of thousands of activists from a wide range of social movements, non-governmental organizations (NGOs) and activist groups, united in common opposition to neo-liberal globalization and all forms of discrimination by the slogan "Another world is possible!". Intended by its founders as an "open meeting place for reflective thinking, democratic debate of ideas, formulation of proposals, free exchange of experiences and interlinking for effective action, by groups and movements of civil society" (World Social Forum 2001, article 1), the WSF has been regarded since its inception as a site for knowledge production. Founded at a historical conjuncture in which the Left was arguably in a state of crisis and fragmentation, but which also had seen the emergence of a multiplicity of movements against neo-liberal globalization, it was conceived as a space in which these diverse currents could come together, engage in dialogue, and begin to elaborate new analyses and alternatives.

One of the most novel – and controversial – features of the WSF has been its supposed status as an "open space". Described by one of its founders as "only a place, basically a horizontal space" (Whitaker 2008, 113), the WSF does not seek to establish consensus around a common set of positions or speak in the name of all participants. It is in principle open to all civil society actors who subscribe to the fairly minimal requirement of opposition to neo-liberal globalization and who are not engaged in armed struggle, and based on the principle of self-organization: those who organize social forums are meant simply to provide a space for participants to organize their own activities (Sen 2010, 997). In this way, the WSF is meant to function as an "incubator" for new initiatives but without itself becoming a political actor (Whitaker 2008, 113).

Central to the notion of "open space" is the rejection of all *pensamientos únicos* (univocal modes of thought) and explicit embrace of plurality. According to Santos (2006, 13–29), the WSF is expressive of an "epistemology of the South": an affirmation of epistemic plurality which seeks to replace the "monocultures" of neo-liberal globalization (and the modern epistemological frameworks that underpin it) with "ecologies" that allow for a multiplicity of knowledges and practices to coexist. On such a reading, the WSF expresses a different logic from the universalizing discourses and grand narratives of the "old Left", and can be seen as a concrete manifestation of an epistemology founded on plurality and irreducible difference.

The "open space" of the WSF undoubtedly has facilitated the convergence of an unprecedented diversity of actors, "creating conditions of possibility for communicative relations across previously unbridged, indeed largely unrecognised differences" (Conway 2011, 219). However, it also has been criticized on a number of grounds. Within the so-called "space versus movement" debate, critics of the open space model have emphasized its inability to foster unified political action and argued for the WSF to become more of a political actor in its own right.[1] Another strand of criticism, meanwhile, has focused on the WSF's failure to live up to its own ideal of openness. Key issues raised in this respect include structural barriers to participation, such as travel costs and visa restrictions (Ylä-Anttila 2005; Doerr 2007; Andretta and Doerr 2007; Vinthagen 2009); the relatively privileged background of the majority of forum participants (IBASE 2006; Santos 2006; Smith et al. 2008); the domination of the WSF by cosmopolitan intellectual elites (Pleyers 2008; Worth and Buckley 2009); exclusions arising from cultural norms that favour conventional modes of political expression (Wright 2005; Ylä-Anttila 2005; Doerr 2007); and the persistence of "global hierarchies of knowledge and power that privilege the modern West" (Conway 2011, 217). Though a self-proclaimed "world process" (World Social Forum 2001, article 3), the WSF is clearly far from global in reach, whether in absolute or qualitative terms.

Cognizant of such exclusions and asymmetries, forum organizers have sought in various ways to "globalize" the WSF in order to bring it closer to grassroots movements around the world. This impulse was behind the decision of the WSF International Council to move the forum from its birthplace in Porto Alegre after 2003, and is also discernible in the multiplicity of social forums that the WSF has spawned on different scales: from continental gatherings in Europe, Africa, Asia and the Americas to neighbourhood-level social forums in cities around the world. A concern to globalize the WSF also has been evident in efforts by activists and organizers to expand or decentralize the world event itself by using new communication technologies to connect actors in different geographical locations. It is this use of communication technologies to construct networks within the WSF process that this article explores.

In what follows, I begin by briefly outlining two attempts at what may be described as "grassrooting" the WSF by bringing it closer to localized actors: the WSF 2008, which in place of a single world event took the form of a Global Day of Action with local activities taking place around the world; and Belém Expanded – an initiative that involved connecting groups in different locations to the WSF 2009 in Belém, Brazil, using videoconference technology. In both cases, it was media and communication that gave these decentralized activities a coherent framework. I then move on to explore in detail an event that provides a different vantage point on the notion of expanding the WSF: the Expanded Social Forum of the Peripheries (Fórum Social Expandido das Periferias), a small social forum held in February 2010 in a poor urban neighbourhood in the southern Brazilian city of Pelotas, which also used videoconference technology to connect activist groups in different places. Through the analytical lens provided by recent scholarship that emphasizes the political and epistemological significance of place in a globalized world, I read efforts by the organizers of this forum to construct a "network of peripheries" through the use of new communication technologies as the expression of a complex politics of place that is simultaneously local and global in scope and which challenges traditional conceptions of scale as well as dominant epistemological paradigms.

"Grassrooting" the WSF: the Global Day of Action 2008 and Belém Expanded

The idea of a completely decentralized WSF was first realized in 2008, when instead of one world event there was a week of mobilization culminating in a Global Day of Action (GDA) on 26 January, with over 1000 activities taking place in 80 countries. Media and communication were integral to the design of the GDA. In order to bring together and give visibility to the numerous activities taking place around the world, a website was created where activists could register and provide information about their actions in designated "spaces" which could be "visited" by others – an initiative which might be described as an attempt to recreate, in virtual form, the physical space usually provided by centralized WSF events. In addition to the website, members of the WSF Communication Commission also coordinated efforts to promote the GDA to international mainstream media, arranged alternative media coverage of the various issues being raised, and organized a set of live connections via Skype, coordinated from France and Catalonia, with activist groups around the world.

The GDA gave rise to the idea that the WSF 2009 also could have a decentralized component in the form of activities taking place simultaneously in other parts of the world, connected in real time to the Belém forum through videoconference technology. During the event, members of the Communication Commission coordinated a programme of activities – brought together under the moniker "Belém Expanded" – which incorporated live interconnections between participants at the forum site and activist groups in other places. Many of these groups had organized their own events in connection with the WSF, including meetings, rallies and performances; these decentralized activities were conceived as part of an "expanded" social forum event encompassing a virtual as well as a physical "territory". During the forum, 30 videoconferences were held with activist groups in different parts of the world, including Europe, North and South America, Africa and the Middle East.

The GDA and Belém Expanded provide interesting examples of efforts to "grassroot" the WSF by bringing it closer to localized actors through innovative use of new communication technologies. Clearly informed by a democratizing impulse, these initiatives provided ways to extend the "forum experience" beyond the world event itself to those who do not have the resources or inclination to travel. In this respect, they might be understood as attempts to realize the WSF's ideals of openness and globality through communication technologies. However, the GDA and Belém Expanded still might be regarded as efforts to decentralize and expand the WSF "from the centre", in the sense that they were initiated and coordinated by actors who occupy relatively central positions within the WSF and took the world event as their spatial and conceptual reference point. The Expanded Social Forum of the Peripheries, by contrast, might be understood as an attempt by actors who occupy a more marginal position within the social forum process to expand the WSF "from the periphery".

Connecting the peripheries: the Expanded Social Forum of the Peripheries

The Expanded Social Forum of the Peripheries was held in February 2010 in Dunas, a poor urban neighbourhood on the outskirts of the city of Pelotas in the southern Brazilian state of Rio Grande do Sul. The event was conceptualized as part of the WSF 2010, which – in accordance with the principle of decentralization – took the form of a series of local, regional and thematic social forums taking place around the world throughout the year. Like Belém Expanded, the Expanded Social Forum of the Peripheries made use of videoconference technology to enable real-time audiovisual interconnections with groups in other geographical locations. Yet, as a social forum that differed both in qualitative and quantitative terms from the biennial world event, it provides a very different perspective on the idea of expanding the WSF.

Situated three hours by bus from the WSF's birthplace in Porto Alegre, Dunas is home to a predominantly Afro-Brazilian population of around 30,000. The neighbourhood suffers from problems that are common to Brazilian favelas: lack of basic infrastructure, low education levels, drug and alcohol addiction, and – not least – stigmatization in mainstream public opinion as a place of violence and lawlessness. However, Dunas also has had some infrastructure put in place in recent years, thanks in most part to the efforts of a well-organized community sector. The local community association, the Dunas Development Committee (Comité de Desenvolvimento Dunas, or CDD), which brings together a number of organizations operating in the neighbourhood, received financial support from a federal government project that enabled the construction of a community centre in 2006. This is home to a small library, a cluster of computers for internet access, multimedia facilities and meeting rooms. Adjacent are a sports stadium and a row of shops for local businesses, all of which are managed by CDD.

The Expanded Social Forum of the Peripheries was held in and around the community centre and incorporated a range of activities, including a solidarity economy fair, cultural and sports activities, and a children's forum – as well as seminars and debates on a range of issues. An initiative of the University of the Periphery (Universidade da Periferia), a grassroots education network that incorporates CDD and various other organizations working in Dunas and nearby areas, it was the latest in a series of social forums held in the neighbourhood over the previous decade.[2] Conceiving the event as part of a global social forum process, organizers adopted the concept of an expanded social forum from the WSF 2009 and made similar use of communication technologies. The majority of activities that formed part of the Expanded Social Forum of the Peripheries were filmed and streamed live online, and many of the seminars incorporated live dialogues with activists in other parts of the world – including France, Spain, Colombia, Mexico and the Amazon – using Skype video call and chat.

As its name suggests, the Expanded Social Forum of the Peripheries had as a key objective to connect different "peripheries", the notion of the periphery being used to refer not only to geographical location, but also to the condition of being marginalized and excluded.[3] Identifying Dunas as being on the periphery, the organizers aimed to establish and strengthen connections with other actors in analogous positions: from similar neighbourhoods in Pelotas to indigenous communities in the Amazon and housing rights activists from the *banlieus* of Paris. The rationale behind this was outlined by one organizer, a man in his fifties from the University of the Periphery, in the following terms:

> We understand that it is necessary to act locally. But it's no use acting locally without a universal vision, without a vision of everything. And you cannot have a vision of everything without seeking articulations with other places, and exchanging experiences between different places. (Interview with the author)[4]

Communication technologies are central to such forms of exchange. The use of video-conference technology to facilitate live audiovisual interactions between activists in different places was conceived explicitly by the same person as a means to facilitate bottom-up processes of convergence between different place-based knowledges:

> When using technologies for sharing of information, for sharing of knowledges that are developed in different places but which in many cases arise from very similar necessities, these knowledges can be shared and transferred and reappropriated by communities in various parts of the world. And this communication makes possible a synthesis of knowledges which are worked out and developed in different regions, within different cultures. (Interview with the author)

Emphasizing the importance of networking with other localities, these excerpts echo Castells's contention that the ability of social movements to create or influence global communication networks is crucial to their success. Observing that in the network society, networks of power are usually global while resistance is usually local, Castells asserts that "[h]ow to reach the global from the local, through networking with other localities – how to 'grassroot' the space of flows – becomes the key strategic question for the social movements of our age" (2009, 52). Like networks of power, alternative projects must also go through global communication networks to transform consciousness if they wish to effect social change: "it is only by acting on global discourses through the global communication networks that they can affect power relationships in the global networks that structure all societies" (53).

Seemingly heeding Castells's imperative to "go global", the organizers of the Expanded Forum of the Peripheries recognize the need to construct communication networks for knowledge exchange beyond their particular locality. However, they also see the politics of communication as inextricably linked to the politics of place. Their networking strategies are intimately bound up with place-*making*; that is, with attempts to construct a particular sense of what Dunas is like as a place. For forum organizers in Dunas, the use of communication technologies is not simply about connecting already existing places with already formed knowledges, nor is it a matter of merely disseminating knowledge through disembodied global communication networks.

The political and epistemological significance of place

The significance of their efforts to "connect the peripheries" through communication technologies can be elucidated by adopting an analytical perspective that emphasizes the political and epistemological significance of place in a globalized world. As Escobar argues, place often has been marginalized in debates about globalization, which have tended to equate the global with "space, capital, and the capacity to transform while the local is associated with place, labor, tradition, and hence with what will inevitably give way to more powerful forces" (2008, 30). Within such frameworks, "local" movements are frequently reduced to, at best, misguided struggles to defend traditional ways of life against modernizing forces, or, at worst, anti-modern fundamentalisms. In contrast to this privileging of the global, Escobar develops an understanding of place and the politics of place that many movements engage in as "key to our understanding of globalization" (15). This politics of place relies on place-making – cultural-political practices concerned with the production of meaning about a particular geographical territory – as a strategy for the defence of place against the delocalizing effects of global capital, but cannot be reduced to mere "resistance" to global forces.

Escobar (2007a) describes the struggles of many contemporary movements as place-based yet transnationalized, involving both the defence of local models of social life and mobilizations involving the construction of coalitions at different geopolitical scales. Osterweil conceptualizes this emergent politics as "place-based globalism" and contrasts it to a "universalizing globalist" perspective, according to which "effective resistance to neo-liberal capitalist globalization must come in the form of a united global movement that has moved *beyond* place-based and local struggles to occupy and constitute *an* alternative global space" (2005, 25). Place-based globalism, by contrast,

> sees true or qualitative globality as comprised of many nodes, places, interconnections and relations that at no point are totally consolidated into a singular global entity. Instead, in their diffuseness and local rootedness they touch and involve increasingly more parts of the globe. (26)

In such a perspective, the place-based character of many contemporary movements does not have to equal insularity or backwardness. Rather, it might be conceived in terms of a positive project concerned with the construction of alternative political and epistemological imaginaries: "an expanding politics of diversity and recognition that acknowledges the multiplicity of alternative visions, values and world views, and the presence of existing 'other worlds'" (Conway 2008, 223). The practices of such movements involve the production of knowledge that is "embedded in locality and that is responsive and accountable to place-based constituencies – as opposed to the detached expert knowledge of modernity" (Escobar 2007a, 286). This can be understood as what Santos refers to as "postmodern knowledge": "knowledge about the conditions of possibility of human action projected into the world from local time-spaces" (2007, 36). Such a perspective highlights the centrality of place – understood both as a particular geographical territory and people's culturally and historically informed experience of, and engagement with, this territory – to the elaboration of alternative knowledge projects.

Conceptualized in epistemological terms, place thus becomes central to any understanding of what "knowledge from below" might mean in a globalized world. This can be elaborated with reference to the notion of the "colonial difference" associated with what is commonly referred to as the Latin American modernity/coloniality programme.[5] According to Mignolo (2000), the colonial difference refers to the space at the exteriority of the "modern/colonial world system"; "the space where *local* histories inventing and implementing global designs meet *local* histories, the space in which global designs have to be adapted, adopted, rejected, integrated, or ignored" (ix). Mignolo understands the colonial difference as a privileged site for the articulation of alternative knowledge projects – "the space where the restitution of subaltern knowledges is taking place and where border thinking is emerging" (ix). Considering the prospect of new macro-narratives, based not on the quest for a counterpart to universal history or an alternative truth but on the search for a different logic, he envisages "an alternative to totality conceived as a network of local histories and multiple local hegemonies" (22).

Communication and the politics of place in Dunas

Within such a framework, the communication practices of forum organizers in Dunas might be regarded as an attempt to speak from the colonial difference and construct alternative global imaginaries. The starting point for their project of "connecting the peripheries" through communication networks is a commitment to Dunas as a place: the majority have strong connections to the area, either as residents or as members of small organizations with long-term involvement in the community. Based on a sense that social transformation has to be grounded in the experiences of people on the ground rather than imposed from above, organizers place a strong emphasis on valorizing local knowledges and practices, and their key objectives are to empower the local resident population and strengthen its capacity for autonomous organization.

Within this schema, the construction of communication networks is motivated by a concern to create a *sense* of globality, which can in turn be mobilized to give impetus to a place-based project of social transformation. A major challenge facing community organizers in Dunas is deep-seated internalized prejudice among local residents about the area in which they live. Dunas is represented in almost wholly negative terms in the local mainstream media, and, with little access to alternative discourses, residents have few resources for constructing more positive self-representations. An important part of the rationale behind the Expanded Social Forum of the Peripheries was therefore to raise self-esteem and stimulate local residents to participate actively in collective efforts to improve the area. Incorporating videoconferences with activists in other parts of the world into what was otherwise a fairly community-oriented event was conceived as an important part of this strategy, as one of the technical coordinators of the event, a man in his thirties, explained:

> If the community realizes that it is […] being seen, it is being looked at, that it is being visited by outsiders, the community has a tendency to like this more, to like and then care for and participate. So […] at the basis of the proposal is this: to make people see that there are people from the outside coming here, to participate, to interact with us here. So it's not such a wretched place, it's not that bad living here. It's not that I don't want a better place to live, but it is better if everybody joins together and works to improve this place here, instead of abandoning it in favour of another place. (Interview with the author)

Complex dynamics are at play here. On the one hand, it would appear that the capacity of the local community for transformative action depends on a validating gaze from the "centre", brought by the physical and virtual presence of international participants (including myself as researcher). Arguably, organizers' deployment of "the periphery" as a political category might inadvertently reinscribe the marginality of Dunas, insofar as their project of connecting the peripheries remains tied to a conventional centre–periphery model. On the other hand, their efforts to create a sense of Dunas as a place that is of interest to "outsiders" and connected to other places through communication technologies is in important ways about staking a claim for Dunas to be situated *in* the world and not simply relegated to the status of the local and marginal, as is usually the case. This sentiment is reflected in one of the main slogans of the forum – "Dunas Mundo no Mundo Dunas" – an approximate English translation of which might be "Dunas in the world, the world in Dunas". The technical coordinator quoted above explained the slogan in the following terms:

> It's this connection to … it's more in the other sense, of bringing the world inside, but not necessarily the outside world. It is to transform Dunas in the world, in its own world, with its own life that … ventures outside, which shares with this other outside world. It is also about bringing this world [to Dunas] but not to live according to this world. It is about generating conditions in which we can guide this outside world, and not have the outside world tell Dunas how it should behave. (Interview with the author)

Part of the purpose of the Expanded Social Forum of the Peripheries, then, was to construct a sense of Dunas being *part* of the global, not just a locality that is impacted upon by global forces originating elsewhere. Significantly, the organizers' use of communication technologies to connect with people in other parts of the world was about creating a conception of Dunas as a place from which knowledge emanates. For a few days, the forum inverted conventional notions of centre and periphery, temporarily placing Dunas at the centre of the world. It was Dunas – not the global WSF event – that was "expanded". This notion was invoked explicitly by one organizer who got up on stage during the forum's closing event to announce that it was broadcast live online and exclaim that "tonight, Dunas is at the centre of the world!"

In the longer term, the Expanded Social Forum of the Peripheries forms part of ongoing efforts to create and strengthen communication networks between Dunas and other "peripheries". The forum provided an occasion for reaffirming already existing relationships as well as for establishing new links, and – not least – to stimulate ongoing dialogue and the formulation of joint strategies beyond the event itself. However, the *sense* of globality and connectedness invoked by the use of videoconference technology was just as important as the actual connections that were made and the content that was exchanged within them. The organizers' innovative use of new communication technologies to create an understanding of Dunas as an important node in global networks is connected to broader efforts to increase self-esteem and encourage a sense of protagonism among the local population. A sense of belonging to the global thus becomes an important resource for a project for social transformation that takes Dunas as its focal point.

The process of social transformation envisaged by the organizers is grounded in place-based knowledges and practices, developed by and for the local population and starting from its particular needs and experiences. Practising a prefigurative politics – "modes of organisation that deliberately demonstrate the world you want to create"

(Grubacic 2004, 37) – they understand knowledge production as inextricably bound up with efforts to implement alternative modes of social organization. One organizer, a man in his thirties from Amiz, an organization set up by students at the Federal University of Pelotas for working with the community in Dunas, gave the following example:

> We can set up a clothes manufacturing business here which doesn't have a boss who deci-des, who is going to exploit people. We can set up an enterprise where people are responsi-ble even when there is nobody who tells them what to do, which is part of a dialogic process. This is a form of truth, a way of knowing differently, of thinking that "yes, I can", and starting from ourselves here in Dunas, begin to think that we can look after the neighbourhood, that we don't have to wait for the public authorities. (Interview with the author)

An idea of truth as produced through practice, through actively *creating* social reality, is at work here. This truth-making is at the same time place-making, focused on the locality as the particular site in which social transformation is effected.

Another slogan used by forum organizers captured this well. A play on the WSF slogan, the Expanded Social Forum of the Peripheries asserted that "Another world is *here*!". The organizer from Amiz explained the thinking behind this in the following terms:

> The "here" is […] the idea that, yes, another world is possible, but where is it that it is happening? It is happening here. So "another world is here" is the answer for us, for our place. So it is here that we are going to act, where we reside, where we live, where we love. It is here that we make the transformation, here that is the other possible world. It's not there. It is here, where we are. (Interview with the author)

The implications of this are twofold. First, it highlights the primacy of place; the notion that social transformation is not an abstract process that occurs elsewhere. The assertion that "another world is *here*" makes it clear that social transformation has to *start from* concrete local realities and practices. Second, it suggests that another world is *already* here; that the kind of social relations that organizers wish to construct already exist – albeit in embryonic form – in Dunas. As the president of CDD, a young woman in her twenties from Dunas, explained:

> We started to think, "ah, another world is here", right? Because Dunas […] is a privileged neighbourhood, various cool things happen here, we have various committed people […]. You could see that the community is a poor community but it is a joyful community, right? Nothing happened, we had four days of the forum and we had no problems. The kids turned up, we can't exclude anyone in the process, regardless of who they are. So it is because of this that another world is here, because the situation is different here, the move-ment is different; the movement is one of inclusion. This is why it is "another world is here". (Interview with the author)

Conclusion: reconceptualizing networks, place and scale

The emphasis that organizers of the Expanded Social Forum of the Peripheries put on place highlights that what is at stake in their use of communication technologies is the creation of networks between different place-based actors and the construction of a sense of globality which does not entail abandoning a commitment to place. It is not about the construction of disembodied global networks that exist *above* particular places, nor an imaginary in which the global is privileged at the expense of the local. Rather, the global – constituted in and through translocal connections achieved through innovative use of new communication technologies – becomes a resource for empowering local struggles.

This place-based yet global politics challenges conventional understandings of place and scale, in which the local is conceived as physically bounded and nested within hierarchies of scale. By seeking – through the use of new communication technologies – to establish translocal connections with other place-based actors engaged in similar struggles, the forum organizers practise what Sassen refers to as an "emergent global politics" that is "global through the knowing multiplication of local practices" (2006, 375). As the notion of place-based globalism (Osterweil 2005) suggests, the idea of globality need not only refer to phenomena that are self-evidently global in scale; practices like the ones described here might also be considered global in that they involve efforts to insert a particular locality in global social and political processes through the creation of transnational networks with actors in analogous positions (Sassen 2007). Through their use of communication technologies, community organizers in Dunas seek to generate a sense of participation in struggles that are globally distributed. Though they remain focused on their particular place, they explicitly frame their struggles as similar to those of multiple other communities around the world and seek to connect with such communities.

Such a reconceptualization of place and scale has implications for the epistemological paradigms through which we understand the knowledge production that these activists seek to facilitate. Within a conventional framework, it might be perceived as local and particularistic, as opposed to more universal and disembodied "global" knowledge. However, the practices described here challenge such a rigid dichotomy between the particular (local) and the universal (global) as well as the hierarchy between them. The organizers of the Expanded Social Forum of the Peripheries want to empower the production of knowledge that is place-based, but not place-*bound*. By using communication technologies to construct networks in which local knowledges can be shared, they seek to enable the production of "postmodern knowledge" projected into the world from local time-spaces, to use Santos's terminology.

The practices described in this article demonstrate what the process of constructing alternative epistemological imaginaries along the lines envisaged by Santos and Mignolo might look like in practice, and the fundamental contribution of communication technologies to such a project. However, the construction of communication networks is about more than connecting already existing places and facilitating exchange of already formed local knowledges between them. As we have seen, a particular sense of Dunas as a place that is part of the global is achieved through the construction of communication networks in which it is an important node. This in turn has the potential to empower autonomous knowledge production and give impetus to a process of social transformation carried out by and for the local community. Through the use of communication technologies, organizers seek to make visible and validate subalternized knowledges, and to inspire among residents a sense of protagonism and capacity to effect social transformation. Their efforts to simultaneously bring the world to Dunas *and* stake a claim for Dunas to be positioned *in* the world illustrate the complex dynamics involved in carving out a locus of enunciation for a community that has been marginalized by hegemonic globalization.

The deployment of the category of "the periphery" by organizers of the Expanded Social Forum of the Peripheries might, on the one hand, function to reinscribe the community's marginality. On the other hand, their concern to "expand" their own social forum rather than simply seek inclusion in the WSF arguably also functions to decentre the biennial world event itself. At least, it raises profound questions about how the "open space" of the WSF is to be conceptualized, and holds out the possibility that "expanding" the social forum process is as much about facilitating the proliferation of autonomous knowledge projects among place-based actors around the world as it is about enabling their inclusion in the global event. The use of communication technologies to "connect the peripheries" clearly is fundamental to such a project, as a means to facilitate information exchange through global communication networks, but also as a way to construct an alternative sense of globality that can give impetus to place-based struggles.

Acknowledgements

The author would like to thank the organizers of the Expanded Social Forum of the Peripheries for their hospitality and acknowledge their intellectual contributions to this article.

Funding

This article is based on ethnographic research which was funded by an ESRC doctoral studentship and a travel grant from the University of London Central Research Fund.

Notes

1. For key arguments in this debate, see, for example, Patomaki and Teivainen (2004), Teivainen (2004), Wallerstein (2004), Whitaker (2008), as well as contributions to the 2005 special issue on the WSF of the *International Journal of Urban and Regional Research* 29 (2).
2. Dunas hosted its first social forum in late 2000, inspired by the preparations for the inaugural WSF in nearby Porto Alegre, which was held in January 2001. Forums held in Dunas since then include the Dunas Social Forum (Fórum Social Dunas) in 2006, the Social Forum of the Communities of Rio Grande (Fórum Social das Comunidades do Rio Grande) in 2007 and the Social Forum of the Periphery (Fórum Social da Periferia) in 2008.
3. The notion of "the periphery" has a particular meaning in the Brazilian context. In general usage, it refers to areas located on the outskirts of big cities and is loaded with connotations of deprivation and poverty. "The periphery" is also claimed as a political identity by many urban social movements wishing to redefine the concept and condition of being on the margins in positive terms.
4. All quotations from interviews have been translated from Portuguese by the author.
5. The modernity/coloniality programme is associated primarily with the work of Argentine/Mexican philosopher Enrique Dussel, Peruvian sociologist Ánibal Quijano and Argentine/US cultural theorist Walter Mignolo. Coloniality, in this framework, refers to the "underside" of modernity – "those subaltern knowledges and cultural practices world-wide that modernity itself shunned, suppressed, made invisible and disqualified" (Escobar 2004, 210) – which has existed alongside modernity since the conquest of the Americas and is, fundamentally, constitutive of it (Mignolo 2000). See Escobar (2007b) for a critical overview.

References

Andretta, Massimiliano, and Nicole Doerr. 2007. "Imagining Europe: Internal and External Non-State Actors at the European Crossroads." *European Foreign Affairs Review* 12: 385–400.

Castells, Manuel. 2009. *Communication Power.* Oxford: Oxford University Press.

Conway, Janet. 2008. "Geographies of Transnational Feminisms: The Politics of Place and Scale in the World March of Women." *Social Politics: International Studies in Gender, State and Society* 15 (2): 207–231. doi: 10.1093/sp/jxn010.

Conway, Janet. 2011. "Cosmopolitan or Colonial? The World Social Forum as 'Contact Zone'." *Third World Quarterly* 32 (2): 217–236. doi: 10.1080/01436597.2011.560466.

Doerr, Nicole. 2007. "Is 'Another' Public Sphere Actually Possible? The Case of 'Women Without' in the European Social Forum Process as a Critical Test for Deliberative Democracy" *Journal of International Women's Studies* 8 (3): 71–87.

Escobar, Arturo. 2004. "Beyond the Third World: Imperial Globality, Global Coloniality and Anti-Globalisation Social Movements." *Third World Quarterly* 25 (1): 207–230. doi: 10.1080/0143659042000185417.

Escobar, Arturo. 2007a. "Actors, Networks and the New Knowledge Producers: Social Movements and the Paradigmatic Transition in the Sciences." In *Cognitive Justice in a Global World: Prudent Knowledges for a Decent Life*, edited by Boaventura de Sousa Santos, 273–294. Plymouth: Lexington Books.

Escobar, Arturo. 2007b. "Worlds and Knowledges Otherwise." *Cultural Studies* 21 (2): 179–210. doi: 10.1080/09502380601162506.

Escobar, Arturo. 2008. *Territories of Difference: Place, Movements, Life, Redes.* Durham, NC: Duke University Press.

Grubacic, Andrej. 2004. "Towards Another Anarchism." In *World Social Forum: Challenging Empires*, edited by Jai Sen, Jai Sen, Anita Anand, Arturo Escobar, and Peter Waterman, 25–43. New Delhi: The Viveka Foundation.

IBASE. 2006. "World Social Forum: An X-ray of Participation in the 2005 Forum: Elements for Debate." IBASE 02 Mar 2012. Accessed 5 September 2013. http://www.ibase.org.br/userimages/relatorio_fsm2005_INGLES2.pdf.

Mignolo, Walter D. 2000. *Local Histories/Global Designs: Coloniality, Subaltern Knowledges, and Border Thinking.* Princeton: Princeton University Press.

Osterweil, Michal. 2005. "Place-Based Globalism: Theorizing the Global Justice Movement." *Development* 48 (2): 23–28. doi: 10.1057/palgrave.development.1100132.

Patomaki, Heikki, and Teivo Teivainen. 2004. "The World Social Forum: An Open Space or a Movement of Movements?" *Theory Culture & Society* 21 (6): 145–154. doi: 10.1177/0263276404047421.

Pleyers, Geoffrey. 2008. "The World Social Forum: A Globalisation From Below?" *Societies without Borders* 3: 71–89. doi: 10.1163/187219108X256217.

Santos, Boaventura de Sousa. 2006. *The Rise of the Global Left: The World Social Forum and Beyond.* London: Zed Books.

Santos, Boaventura de Sousa. 2007. "A Discourse on the Sciences." In *Cognitive Justice in a Global World*, edited by Boaventura de Sousa Santos, 13–43. Plymouth: Lexington Books.

Sassen, Saskia. 2006. *Territory, Authority, Rights: From Medieval to Global Assemblages.* Princeton: Princeton University Press.

Sassen, Saskia, ed. 2007. *Deciphering the Global: Its Scales, Spaces and Subjects.* London: Routledge.

Sen, Jai. 2010. "On Open Space: Explorations towards a Vocabulary of a More Open Politics." *Antipode* 42 (4): 994–1018. doi: 10.1111/j.1467-8330.2010.00785.x.

Smith, Jackie, Marina Karides, Marc Becker, Dorval Brunelle, Christopher Chase-Dunn, Donatella della Porta, Rosalba Icaza Garza, Jeffrey S. Juris, Lorenzo Mosca, et al. 2008. *Global Democracy and the World Social Forums*. London: Paradigm Publishers.

Teivainen, Teivo. 2004. "The World Social Forum: Arena or Actor?" In *World Social Forum: Challenging Empires*, edited by Jai Sen, Jai Sen, Anita Anand, Arturo Escobar, and Peter Waterman, 122–129. New Delhi: The Viveka Foundation.

Vinthagen, Stellan. 2009. "Is the World Social Forum a Democratic Global Civil Society?" In *The World and Us Social Forums: A Better World is Possible and Necessary*, edited by Judith Blau and Marina Karides, 131–148. Plymouth: Lexington Books.

Wallerstein, Immanuel. 2004. "The Dilemmas of Open Space: The Future of the WSF." *International Social Science Journal* 56 (182): 629–637.

Whitaker, Chico. 2008. "The World Social Forum as Open Space." In *World Social Forum: Challenging Empires*, 2nd ed., edited by Jai Sen, Anita Anand, Arturo Escobar, and Peter Waterman. London: Black Rose Books.

World Social Forum. 2001. "Charter of Principles." *Fórum social mundial*. WSF. Accessed 5 September 2013. http://www.forumsocialmundial.org.br/main.php?id_menu=4&cd_language=2.

Worth, Owen, and Karen Buckley. 2009. "The World Social Forum: Postmodern Prince or Court Jester?" *Third World Quarterly* 30 (4): 649–661. doi: 10.1080/01436590902867003.

Wright, Colin. 2005. "Opening Spaces: Power, Participation and Plural Democracy at the World Social Forum." *Ephemera* 5 (2): 409–422.

Ylä-Anttila, Tuomas. 2005. "The World Social Forum and the Globalization of Social Movements and Public Spheres." *Ephemera* 5 (2): 423–442.

Arguing about religion: BBC World Service Internet forums as sites of postcolonial encounter

David Herbert[a], Tracey Black[b] and Ramy Aly[c]

[a]University of Agder, Norway; [b]University College London, UK; [c]University of Sussex, UK

This paper compares the content, context and facilitation of discussions about religion and politics on contrasting types of Internet forum in English and Arabic, comparing forums linked to the current affairs output of the BBC World Service (BBCWS), Al-Jazeera and Al-Arabiyya between 2007 and 2010. It examines how these forums are structured by a range of technological and political factors, reflecting the postcolonial context of encounter. It demonstrates that the consequences of deploying these technologies in this context are ambivalent. On the one hand, they enable participation in discussion in contexts where political debate is restricted, allow conversation between individuals that would be impossible in embodied form, and strengthen diasporic connectivity. On the other, they are used behind a rhetoric of free speech which masks exclusionary power relations. Comparison between types of BBC forum further demonstrates the importance of how new technologies are deployed for forms of communication and community construction.

[I]f the adherents of a religion enter the public sphere, can their entry leave the pre-existing discursive structure intact? The public sphere is not an empty space for carrying out debates. It is constituted by the sensibilities – memories and aspirations, fears and hopes, of speakers and listeners, and also by the way they exist (and are made to exist) for each other.

(Asad 1999, 181)

Asad's comments are made in the context of broad reflections on religion, state and secularism, but they also provide a useful way of approaching the online forums that are the focus of this article. "Have Your Say" forums may be presented as "empty spaces" for the exercise of free speech (Hain 2000; Foreign and Commonwealth Office [FCO] 2006). However, they are moderated to ensure compliance with house rules; the topics discussed are selected by editors, or through some kind of blog dialogue; discussions are seeded by carefully crafted comments and media feeds in a particular international news environment, and in some cases debates are shaped in their appearance through user recommendation. Furthermore, behind the existence of the forums there is also the entire political narrative of how BBC forums come to be funded, which includes relationships between government, the FCO and the BBC World Service (BBCWS), through which World Service targets are set and performance assessed, and between different branches of the BBC, especially the World Service and bbcnews.com, which shape the flow of content across BBC platforms, including blogs and forums. In these flows and relationships, Asad's broader concerns about the articulation between the mutually defined categories of religion, state and secularism also surface, since each of these levels contributes to shaping how forum users and viewers "exist and are made to exist for each other".

Nevertheless, contributors are not simply positioned by the framing discourses and production processes which structure forums. Rather, they actively react to the framing provided by the service editors, challenging the formulation of questions, the choice of debate-seeding material, and of course each other. In these ways, forums provide opportunities for practices of transculturation, or the transformation of dominant discourses through their reuse by marginalized people. As Bakhtin writes: "The word in language is half someone else's. It becomes half one's own when the speaker populates it with his own intention, her own accent" (1981, 293). In addition, the same digital technologies and editorial pathways which facilitate the centralization of news production centralization also enable media contraflows: that is, flows of media content from marginal to dominant areas of media production (Cottle 2006, 162–164). In particular the media practices used by the *World Have Your Say* (*WHYS*) team provide opportunities to examine examples of "contraflow" in practice. Furthermore, while these multiple language forums may seem a marginal phenomenon, participants and audiences are globally dispersed and part of a service which, despite recent cuts to certain language services, operates around Internet access and restrictions on BBC broadcasting in some countries (e.g. Iran and China), and so enable "conversation" between a potentially vast range of participants; this is, given BBC's weekly reach of 230 million people, around 2 percent of the world's population.

This article raises two sets of broad research questions focused on these forums and offers some tentative answers. On the one hand, what do these forums tell us about the relationship between technology and the postcolonial? Thus, to what extent, if any, do the colonial history, metropolitan base and production relations embedded in a contract with the FCO shape the platforms that the BBCWS offers, and the discourses produced on them? On the other, do the forums represent at least a partial realization of the utopian hopes of some early web developers, of enabling the many to talk to the many across boundaries of geography, power and wealth? Do they provide spaces for the politically oppressed or marginalized to express and share their views? In particular, do they provide any significant contraflow to what Dahlberg (2005, 160) has called "the corporate colonization of cyberspace" and do they in any way support or help to "deepen" democracy (Heller 2009, 125), as some BBC and FCO rhetoric claims? In general, do the multiple public spheres or "spherecules" that they provide enable new forms of interaction, leading to a more nuanced public discussion than established news formats produce? Finally, if so, what factors in forum design and media practice best facilitate such discussion (Wright and Street 2007; Jones and Rafaeli 2000)?

The research springs from an international collaborative project examining the relationship between the BBCWS and its diaspora audiences and contributors, historically and in the present. It also draws on a strand of the project which examines how talk about religion and politics is facilitated on and around the BBCWS current affairs programming. Here, we shall focus in particular on two kinds of web-based forum supported by the BBC in connection with news and current affairs programming: one supported by the World Service directly, and the other by the BBC's news division: bbcnews.com. These two forums differ with respect to discussion sourcing, moderation practices, inter-platform dynamics and embedded news cultures. This enables analysis of how such factors interact to shape talk about religion, and reflection on whether and to what extent these differences influence their functioning as some kind of public sphere, actually or potentially contributing to the practice of democracy (as is implied by some BBC and UK government rhetoric supporting their establishment).

Why religion? Public religion as a contentious global discussion topic

The consensus on the inevitability of secularization – in the sense of the declining public role of religion – amongst western publics (Habermas 2008) and western-influenced elites elsewhere (Rajagopal 2001) has broken down in recent years. It is useful here to distinguish between the public role or social significance of religion on one hand, and patterns of religious observance and cultural transmission on the other. These are distinct, although there is a resource relationship between them. That is, religion cannot be publicly mobilized unless its symbols and discourses have some resonance, and religious symbols will lack that resonance unless their meanings have been successfully conveyed across generations; in this process, observance plays a key role. Insofar as there is a new consensus, it is along the lines that while traditional forms of religious observance continue to decline amongst western European populations and in many societies where people of Western European origin constitute a majority (Norris and Ingelhart 2004), this is not the case amongst other groups, and it is also true of large subcultures in the US. In particular, the assertion of religious identity, and the public relevance of religion in a number of ways, has become if anything more prominent – not least in western Europe, where largely secularized majorities coexist with migrant groups for whom religion often plays a larger role. This condition Habermas describes as "post-secular" (2008, 19), in the sense that secularism and religion must learn to live with one another in contexts often marked by postcolonial dynamics.

As a result of these developments, the public role of religion has become an increasingly contested topic, and one on which attitudes vary markedly both across populations and globally. Hence talk about religion in a current affairs context provides a good test case for assessing the capacity of online forums to sustain debate between globally dispersed contributors with often contrasting views, and for examining the BBC's role in hosting such debates.

Why the BBC World Service? BBCWS as a postcolonial institution

The World Service is an institution which has exhibited a range of responses to negotiating historical transitions, such as from colonial to postcolonial, or the complexities of the Cold War and its aftermath. It was founded in 1932 as the Empire Service for "people who regard Britain as their home", but by 1946 the director general could declare that our "field is now the world", reflecting a recognition of the importance of diverse audiences across the Empire and Commonwealth, not least because of their part in the war effort, and the role of the service in galvanizing this. In 1946 it became the External Service of the BBC, divided into European and Overseas divisions, reflecting Cold War priorities, while the term "World Service" was first used for the English language Overseas Service in 1965, reflecting African decolonization. Since the Cold War it has refocused and reorganized. Ten language services have been cut, mostly in the eastern European area, reducing the total number from 43 to 33; resources have instead been focused on the Middle East, and to a lesser extent China, with BBC World Arabic and Persian TV channels launched in 2007 and 2009.

This altered shape suggests further why the BBCWS is significant for understanding global media dynamics. In spite of an increasingly competitive and commercial media environment, it remains a major global player, reaching (in 2007) 230 million weekly users – approximately 2 percent of the world's population – on a regular basis through its multiple platforms (BBC Trust n.d). It is also a peculiar hybrid. While funded by the British state and associated with "broadcasting Britishness", it also curiously retains a substantial (though, of course, not uncontested) reputation for journalistic impartiality and widespread local coverage. The service was at the time of the research (2007–10) funded by a government grant-in-aid administered by the FCO, with monthly meetings between FCO and WS managers to monitor the Broadcasting Agreement between them. Thus, while the WS has managerial and editorial independence, the Broadcasting Agreement is designed to ensure that the service meets objectives in line with "strategic international priorities" set by the UK government – hence the recent switch in the prioritization of language services.

Lately these objectives have included identifying priority target audiences (by segment and region, such as opinion-formers in the Middle East) and the development of online services. Some of the thinking behind the development of online services, including forums appears in the following extract from an FCO White Paper:

> With the spread of satellite television and the internet, governments are increasingly influenced by public opinion, both domestic and from abroad. A wide range of organisations, some funded by the Government such as the British Council or BBC World Service, some with no links to Government, are actively involved in steering this debate in favour of the values of tolerance, free speech, respect for human rights and democracy. (FCO 2006)

The "global conversation", the contested concept of the public sphere and its methodological implications

A close connection between FCO and BBCWS thinking can be seen in the idea of a "global conversation", found on both the FCO and WS websites. This idea, however, appears to rest on a conception of the public sphere which has been widely criticized since Habermas's influential work on the term in the 1960s, not least from postcolonial and feminist perspectives. Philosopher Charles Taylor exemplifies a conventional use of the term as:

> a common space in which members of society a deemed to meet through a variety of media: print, electronic and also face to face encounters; to discuss matters of common interest; and thus to be able to form a common mind about these. (2004, 83)

We understand the public sphere as consisting of virtual or real sites in which debate about matters of shared concern can be aired. Since Habermas's influential history ([1962] 1989), the concept of the public sphere has been widely discussed, criticized and reformulated (Fraser 1992; Benhabib 1992), and his abstract concept, based on a culturally limited range of historical examples, has been challenged and largely replaced (Habermas 1996) by understandings which recognize diverse culturally embedded communication contexts as both multiple sites of the present public sphere, and historical precedents shaping its formation (Mayhew 1997; van der Veer 1999; Herbert 2003; Bonham 2004). The value of multiple sites with different purposes has also been stressed in research on online forums (Linaa Jensen, 2003, 372).

This kind of understanding has been criticized because it fails to take seriously the power dynamics at work in existing public spheres. Access to, and mastery of, rhetorical skills amongst participants within such spheres is likely to be unequal. In response, critics have stressed the value of multiple public spheres as a response to problems of inequality of access to mainstream public spheres, and ICT developments have enabled this idea to be developed in two significant directions. One trajectory, particularly inspired by the growth of Internet-based discussion forums, is that of multiple "public spherecules" (Gitlin 1998; Cottle 2006, 51), spaces which connect mainstream public spheres with alternative counter public spheres, such as media outlets serving diasporic audiences or specific lifestyle communities. While primarily functioning as discussion forums between individuals, and hence not usually transferring to the mainstream public sphere in the Habermasian sense, many alternative spheres attract a diversity of participants, at least in terms of geographical distribution, and are even more widely viewed than their mainstream counterparts. They provide a context for a certain kind of public conversation, and may on occasion enable media contraflows from less powerful to more powerful media outlets (Cottle 2006, 162–164), as well as serving more mainstream media interests. This article will consider some of these possibilities.

While Habermas (2006) has raised concerns about the dangers of online communication fragmenting public space (Milioni 2009, 409), others disagree; for example, Mitra argues in the context of Non-Resident Indians (NRIs) in the US:

> a combination of economic, technological and cultural processes have worked together to create an ambience where marginalized immigrants have been silenced within the public sphere of the country of adoption. The consequences of the absence of voice have been brought into sharper focus following the terrorist attacks on America […]. I would argue that the Internet could provide […] a discursive location where the traditionally powerless are able to speak for themselves. (2005, 378)

A second direction in which the multiple public sphere model has been developed is the concept of the "public screen". In articulating this concept, De Luca and Peeples (2002) argue that thinking about the public sphere has been dominated by the idea of the face-to-face conversation as the ideal model of human communication, and the belief that the conduct of politics should somehow approximate as closely as possible to it:

> although the public sphere includes written forms of communication, embodied conversation functions as the ideal baseline. Yet the dream of the public sphere as the engagement of embodied voices, democracy via dialogue […] compels us to see the contemporary landscape of mass communication as a nightmare. (2002, 130)

Instead, drawing on earlier cultural critics such as Derrida and Peters, De Luca and Peeples encourage us to think of communication not just in terms of conversation, but also of dissemination, much of which, as in the parable of the sower, falls on rough ground, but some of which survives and blooms. The dependence on conversation as a root image also means that we miss out on the potential of visual practices to act as forms of cultural critique: as they write, "critique through spectacle, not critique as spectacle" (2002, 135). Criticisms of the conversational model of the public sphere have also been articulated in the public use of reason debate. Gaus and Vallier (2009) argue that this is too tied to a particular, deliberative model – politics as a forum in which outcomes ought to reflect the best argument – because it presupposes the existence of agreed criteria of judgement. By contrast, critics argue that politics may be also imagined as a market in which outcomes are determined by a process of bargaining and in which argument jostles with calculation based on diverse interests. Seen in this way, the fact that some premises from which diverse citizens argue are incommensurable is less problematic; in practice such incommensurability is in any case widespread and differences spring from a range of philosophical and cultural as well as religious bases. In such a system, there is no need to leave specifically religious differences at home.

These perspectives have implications for how the discourse on forums is evaluated. For example, Dahlberg (2001) developed a coding system to assess the level of communicative rationality (Habermas 1987) found in forum exchanges, and we derive some of our categories of communication from this. However, forums may also be used to display solidarity, express outrage and strike rhetorical postures, all of which may contribute to a lively public sphere but do not score highly on communicative rationality. We therefore used a mix of methods to address our research questions, depending on question and forum type, and attempt to represent the diversity of communicative forms present. We studied printouts of forums and transcripts of radio broadcasts for examples of types of interaction. Where we sought to compare forms of communication across forums, we counted the incidences of different types of interaction and contribution, using categories that are outlined below.

The forums

1. World Have Your Say

The English language *World Have Your Say* phone-in show is broadcast each day of the week. It is preceded by an interactive blog which tests and generates ideas for the show's topic, and, from the time research commenced and until 2012,[1] was followed by a forum for ongoing discussion, usually kept open for several days. Text messages are read out on the programme, adding to the forum discussion, enabling the representation, if not dialogical participation, of those lacking online access. This is particularly important in Africa, where half of the WS's regular radio audience lives. This forum is reactively moderated, allowing contributions to be instantaneously posted, increasing the sense of "live conversation". Twitter and Flickr feeds provide further text and video-clip participation. As well as seeding content via the blog, there have been experiments with opening the editorial meeting up to audience members through phone conferencing. Other participative methods include open broadcasts, where the programme is broadcast live from various non-studio locations, and open-microphone events, in which the hosting of the show is handed over to participants on location.

Religion and politics are frequent discussion topics, accounting for 17 percent of 40 percent of debates in the periods examined (April and July 2008; April 2010). Topics included "Should all political parties be secular?" (28 July 2008); "Is it ok for religion to enshrine inequality?" (3 July 2008); "Can Muslims take a joke about Islam?" (3 April 2008); "Does religion stand in the way of science?" (1 April 2008); "Should polygamy be legal?" (18 April 2008); "Should national law incorporate sharia law?" (8 February); "Should all Western countries ban the burqa?" (1 April 2010); "Is the Pope above the Law?" (14 April 2010); and "Are Muslims always under attack?" (22 April 2010).

Discussing a religious case study: "Can Muslims take a joke about Islam?"

On 3 April 2008, *WHYS* editor Ros Atkins started a blog which evolved into that evening's discussion on the topic "Can Muslims take a joke about Islam?" Given its sensitivity and that of dominant negative media representations of Islam (Poole 2002), it is perhaps not surprising that many postings answered emphatically "No". However, the debate also included some criticism about its framing, particularly of how Islam is singled out. It was also peppered with contributions, some from Muslims, pointing out that Muslims do indeed tell jokes about each other and their religion. Arguably, most striking was the presence of Muslim contributors from a range of backgrounds, pre-empting their being branded as humourless stereotypes. Some exchanges exemplify how under such conditions debate can be civil, even if profound disagreements – and misunderstandings – remain:

160 Mustafa August 17, 2008 at 10:37 pm

(1) if u guys had been to any muslim country outside of the middle east, you would realize that many muslims make jokes about themselves.

(2) the reason muslims, especially in the middle east, are against so many jokes is because their image is already been hurt by september 11 and they cant tolerate the fact that so many people have a negative view on them.

(3) some people were talking about protests after the danish cartoon. thats because printing pictures of prophets is AGAINST OUR RELIGION. obviously we're not gonna laugh along with that.

(4) if u want proof with other people, just look at the african american population. it might not be against their religion, but they will be very upset if u call them the "n" word. somebody probably will hit u if u said that to them.

161 dwightofcleveland August 17, 2008 at 11:49 pm

Mustafa,

Let make sure that I understand you correctly, and maybe shed some light on "non-Muslim" logic.

(1) Yes most Muslims are not as sensitive and extreme as the "squeaky wheels" that gets the media attention. However, you never see a bear in the news unless it attacks somebody. I would never take a hike in the upcountry without my side-arm. Though every bear I have ever crossed has not attacked me. I can't know which ones are aggressive, so until I know otherwise I must judge each one with caution. All people see on TV are the angry extremist Muslims. It is a natural reaction in that light to judge each one with caution. This is especially true the more one presents him/ herself in the stereotypical way.

(2) So what you are saying is that Muslims are hurt by being portrayed as malice extremist killers. Furthermore, to counteract that undesirable image, some Muslims see it as logical to call for the death and desecration of non-Muslims? They pray this wish will be granted in "Allah's Name". See the logic here is corrupted. We have one side saying the other side are intolerant killers and the other side saying, "no we are not and if you don't believe us we will kill you."

(3) People are cruel. If they know something bothers you they are going to do it. It gives them power over you. My guess is that those Danish cartoons looked nothing like the image of your prophet. The difference between a belligerent

ideology and a peaceful one is this. In Christianity, if somebody offends you or breaks the laws of "God" you are to forgive them and pray for their forgiveness from God. A belligerent ideology is offended and launches an attack on the person calling for his demise cursing them to Hell. The idea behind the peaceful approach is that it ends the circle of violence.

(4) African Americans call each other the "n" word all the time. It is in their music, culture, and lingo. Even when it is used, or some derogatory statement about them such as "nappy headed ho's", is leveled at them, nobody is calling for the death and dismemberment of the person who said it. At worst they are calling for them to loose their jobs. I don't see Muslims making their own cartoons and calling each other names affectionately.

> All of this doesn't explain the London teach being tried and convicted for calling a stuffed animal the most common name in the Middle East, "Mohammed". A teacher, trying to help your children grow, doing something simple and what should have been fun.

The point here is not to analyse, let alone attempt to settle, the arguments. Instead, it is to highlight the continued conversation between participants with very different backgrounds and opinions – in this case, a large-scale conversation that lasted for several days. Furthermore, such discussions enact in practice the shift in academic discourse (highlighted at the start of this article) towards situated rather than universalist understandings of political conversation. Such patient, if sometimes edgy, dialogue contrasts with dominant news debate forms which tend to polarize debates into bipolar forms (Benson 2009), and news formats, which constrain the scope of debate topics (Altheide 1991).

WHYS provides context by drawing on eyewitness testimony. Furthermore, while some early analysts concluded that online communities will tend in practice to be highly specialized (e.g. Beyers 2004), *WHYS* suggests that extremely diverse virtual communities, nourished by regular voice input, can be sustained over time.

Global conversations & media contraflows

The editorial team's proactive role in promoting user-generated content, means that *WHYS* is the only forum where the creation of media contraflows by influencing mainstream news agendas can be discussed. These examples involve the team's role in breaking new stories, or increasing the focus on stories that were already in low-key circulation. The first was the publication of controversial cartoons depicting the prophet Muhammad by *Jyllands Posten* newspaper in Denmark on 30 September 2005. Mark Sandell (2008) describes the process which provides some insight into the difficulties of breaking into the mainstream news agenda, but also how attitudes within the BBC have changed as *WHYS* became more established:

> our programme was the very first programme to get in on the cartoons because our online community was sending them up to us. [...] We were 2 days ahead of anyone else in this building. [...] The first day we did it [...] they [BBC News] just ignored us basically. They just thought [...] it wasn't really news. [...] They wouldn't do that now because the programme's got much more established. [...] The following day at the meeting, I did say "We're going to have to do it again because it has gone absolutely mental". By the end of the second day, it was starting to run on bulletins and all this kind of thing. It was starting to move.

Other examples include the near contemporaneous coup d'état in Thailand (18 September 2006), and riots in Budapest on the 50th anniversary of the 1956 uprising (19–26 September 2006). In each case a low-key media story was transformed by the WS having listeners phone in directly from the scene. In these cases, *WHYS* listeners become the local antennae of the WS, closer and more sensitive to local conditions than foreign correspondents. However, do such examples provide evidence of anything more than an early warning system for the BBC? Do the voices of those initial amateur reporters survive the flood of professional reporters and expert analysts that will follow if a story takes off? There is also the critical question of who these listener-reporters are. Blog participants are certainly more likely to be men than women, and come from elites within many societies simply because they have access to Internet facilities. Sandell himself expresses awareness of the issue of social class in terms of BBCWS staffing: "everyone talks about how cosmopolitan the World Service is. And it is cosmopolitan, but not in terms of class … Not in any way" (Sandell 2008). Furthermore, the profile of bloggers suggests that the same concern might be raised at this level of user.

In a literal sense, the core phone-in programme is often a global conversation, in which a cast as unlikely as a Moscow-based Somali taxi driver, an Indian merchant seaman in the Bay of Bengal and a Texan farmer discuss American foreign policy in the presence of a live global audience.[2] Both forums and phone-ins feature regular contributors, and over time interaction through voice and text has arguably created a sense of community, as editor Mark Sandell argues:

> [T]he blog certainly […] has built a community to an extent. I would say even more than a community, there are people who are regulars on our blog who, genuinely I would say have, have become friends of people who are on the team. So it's actually gone a stage further. They actually feel they've become a part of it. (Sandell 2008)

What is the significance of this kind of transient virtual community? Can it provide some kind of counterpoint to dominant news discourses? And how widely is such a sense of community found here replicated across broadcast related online forums? The *World Have Your Say* media mix is richest and most interactive and participatory of the forums considered; other BBC forums have more limited opportunities for listener agenda-setting and sometimes employ full moderation when topics are perceived as potentially controversial, so that there are delays in posting of comments in order to reduce the conversational "feel". Nonetheless, if not a sense of community, such "message board forums" do generate strings of responses and global juxtapositions, if not conversations.

2. *Have Your Say*

While linked to the World Service site, BBC *HYS* message boards are edited by bbcnews.com. Hence they are a good example of how the World Service and bbcnews.com sites intersect under a new overarching management structure of the news divisions. They differ from *WHYS* in a number of ways. First, platforms: the forum was partnered by a weekly phone-in television programme on *BBC World* (also called *Have Your Say*), which was broadcast simultaneously on radio and the Internet. While there is a suggestion facility for topics, there is no discursive blog to seed topics, which are, rather, chosen by editors. All postings are checked by moderators before being posted, and only a proportion of comments get posted (between 10 percent and 100 percent, depending on the volume of responses). Being closely linked to the main news site, this forum receives a much higher volume of contributions for the most popular topics – up to 12,000 in our study. This site features a recommendation facility for regular users, who can choose to recommend comments that they like. Because users can sort comments by number of recommendations – as a way to get a quick summary of the debate – we found that the most popular comments were not reflective of the debate as a whole.

We examined 24 debates between January and July 2007, including the topics: "Should the churches be able to opt out of gay rights laws?", "Should the veil be banned in schools?", "What does the election result mean for Turkey?", "Should the UK fund the training of imams?", "Should Shambo the bullock be saved from slaughter?" and "What will constitutional change mean for Egypt?".

Religion and state

A significant proportion of the analysed debates centred on the relationship between religion and politics, and, in particular, religion and the state, with several debates exploring this in depth. "What will constitutional change mean for Egypt?" was prompted by the approval of amendments following the Egyptian referendum on constitutional change, which was boycotted by the Muslim Brotherhood. Posts came from at least 18 countries, with a majority (24 percent) from Egypt. Hence the debate provided a forum in which Egyptian users were able to freely discuss with each other and foreigners at a time when national news media were largely state-controlled, and it provides an insight into the development of the Arab blogosphere before the Arab Spring (2011). Registered users submitted 44 percent of the comments and 78 percent of all comments received at least one recommendation. Only a minority of published contributors supported the amendments, although some Egyptian contributors and a large proportion of non-Egyptians expressed relief at what they viewed as a restriction on official religious influence. For example:

> I think this is an incredible step for the Egyptian people. Imagine not being threatened by religious extremists every time you voted or had a different opinion. Imagine actually living your life free. Look at the religious Death squads, roaming the streets of so many countries-beheading people because of their beliefs or opinions are simply different. WHY? Who gives them this right? NO ONE! Every human being deserves freedom! CONGRATULATIONS EGYPT my favorite country.

> Gregory … , Georgia USA

Such opinions were contested by many contributors who resisted the argument being framed as a choice between a secular and religious state, defending principles of openness and fairness in the workings of government. In this context, the debate also contained criticism of foreign powers viewed as complicit in keeping Mubarak in power, or promulgating a double standard in their attitudes towards the role of religion and the state. Some contributors also suggested that suppressing political opposition (religious or secular) automatically encouraged political violence. The majority of contributors viewed the failure of Mubarak as increasing support for the Muslim Brotherhood, or creating the impression that the Brotherhood enjoyed more support than is actually the case. Even opponents of the Muslim Brotherhood voiced uneasiness about government policy: "it will mean that we Copts will fear less of Islamists but more of the government" (Mina … , Alexandria). Some contributors suggested the emphasis on religion and the Muslim Brotherhood drew attention away from the detail of the constitutional changes and the wider implications they will have for democracy in Egypt:

> PEOPLE! FORGET the ban of religious parties. it is negligible to the other amendments: the removal of judicial supervision in future elections, the new "terrorism" law which replaces the 26-year-old "emergency" law! which incidentally give authority to the police to carry out arrests, search homes, conduct wiretaps and open mail without a warrant. very democratic! it also gives the president the authority to order civilians tried by military courts, with limited rights. very democratic. Sad!

> marwa … , Cairo (Egypt)

This debate illustrates the capacity of the message boards to provide a public sphere of discussion for those living in a context where open debate was restricted. In Egypt, for instance, the press was subject to restrictions and periodic crackdowns, and, shortly after this debate was posted, editors of several Egyptian newspapers were arrested for publishing rumours alleging that Mubarak had died. "What will constitutional change mean for Egypt?" also illustrates the capacity of the format to successfully sustain quite complex debate – a factor which arguably exists in tension with the manner of presentation of the debate prompts, and its news orientation. In this sense it would seem to support the democratization argument put forward by the FCO, although the impact of an English-language forum would have been limited. But what of the much more expensive and recently launched Arabic TV channel and linked online forums? What can we learn from these?

3. *Arabic forums and comparison with English forums*

BBC Arabic: context and competitors

The launch of the BBC's new Arabic-language television channel in April 2008 occurred at a time of unparalleled proliferation of government-funded Arabic-language television broadcasting from within and outside the Arab world. This new battle for the "hearts and minds" of viewers in the Arab world is largely a contest over how the perceived existential threats in the region are described, framed and understood. Here the BBC joins an already crowded field, as Table 1 demonstrates.

To compete in this busy and politically charged news marketplace, the BBC has emphasized what it considers to be the strengths of its current and social affairs services

Table 1.   Arabic-language satellite television channels launched, 1996–2008.

Name	Translation	Sponsor	Launch
Al-Jazeera	The Island (peninsula)	Qatar	1996
Al-Arabiya	The Arabic one	Saudi Arabia	2003
Al-Alam	The World	Iran	2003
Al-Hurra	The Free One	USA	2004
DW Arabic		Germany	2005
France 24/7 Arabic		France	2007
Russia Today		Russia	2007
BBC Arabic TV		UK	2008

compared to that of other Arabic satellite TV producers, stressing its traditions of journalistic impartiality, distinctive content (discussing issues such as social and gender matters not discussed elsewhere) and participative ethos and process. However, our analysis of the reception context and some responses on forums suggest that each of these claims is problematic: notions of journalistic impartiality are regarded with scepticism in a context in which all national and transnational news sources are popularly viewed as "directed media"; a European-based broadcaster's handling of sensitive issues such as gender has attracted controversy and criticism; other providers also offer forms of web-based participation linked to their broadcast output. Furthermore, the language of impartiality misses out on a key dimension of the "battle for hearts and minds" in the Middle East – the role of social memory, loyalty and solidarity in the public sphere of Arab societies, and the need for any broadcaster to resonate with this to attract a popular audience.

Our analysis of Arabic forums focused on coverage of the 60th anniversary of al-Nakba (the Disaster) in May 2008, comparing participatory media provided by BBC and Al-Jazeera Arabic and English services, and the Al-Arabiyya's Arabic Service. BBC Arabic's flagship programme نقطة حوار (Nuktat Hiwar or "Talking Point") was the main platform for discussing the May anniversaries. Indeed, the programme is one of the principal platforms for audience participation on the channel. It operates as a 50-minute TV phone-in discussion which continues on the BBC Arabic radio for a further 40 minutes. Throughout these broadcasts, web users and viewers can write in with their comments via email, text or video, while a member of the programme's team surveys the threads and contributions, providing for the viewers a regular gauge of public opinion. The offer of participation is central to BBC Arabic's marketing strategy, and is probably vital if, as a European-owned broadcaster, it is to establish its credibility amongst Arab audiences. Hence, in addition to the Talking Point programme, BBC Arabic's website contains a number of features eliciting audience participation; including a list of the most popular news items and a number of features calling on viewers to "participate with your opinion", "contact us" or send "your video participations".

The phone-in format is popular among TV audiences and broadcasters across the Arab world, principally because, in a region where the views of the average citizen are not taken seriously and where there are almost no effective mechanisms, platforms or forums where their personal opinions can be aired publicly, the phone-in format provides viewers the rare opportunity to literally "air" their opinions. In its early years, many of Al-Jazeera's programmes built their reputation and following on the fact that normal citizens were given the opportunity to call in to live debates to give their opinions. While Al-Jazeera still offers its viewers many opportunities to participate during live programmes, the number of live phone-ins has decreased dramatically. Instead, viewers are encouraged to write in via email or SMS. Possibly the tendency for phone-ins to attract views highly critical and often abusive of Arab leaders has led Al-Jazeera to opt for modes of participation that the channel can more easily control.

Both BBC Arabic and Al-Jazeera offer viewers, listeners and web users the chance to air their views on open-ended questions within their forums. In both cases, participants must register as forum users and adhere to the rules of participation, at risk of exclusion. These measures work towards creating a community of discussants and to encourage user accountability. However we faced a comparability problem because, surprisingly, despite the availability of an extensive and archived forums section on Al-Jazeera, there were no forums open on the subject of the Nakba during May 2008, so a direct comparison of forum data from Al-Jazeera and BBC Arabic on the topic was not possible. In practice, during May 2008 the overwhelming use of the comment forms on Al-Jazeera sent thanks and praise, with little or no discussion, for the huge resources and attention the channel had given to the commemoration of the Nakba. Given Al-Jazeera's audience and this evidence of unequivocal support for its openly partisan stance, it seems there was little to discuss; however, this was not true of its English forums, as we shall see below.

Al-Arabiya does not have a forums feature on its website, but attracted feedback in reference mainly to translated agency news items or editorials from mostly Gulf newspapers republished on the channel's website. Readers' comments were therefore framed in reference to a particular article or programme and not to an open-ended question for deliberation. In contrast, BBC Arabic does not offer this facility to its readers. A further problem of comparison is differences in moderation policies across the services: language routinely used to describe Israel in the nationalist repertoire on Arabic

Table 2.   Discursive repertoires in Arabic forums.

المرجع القومي	The nationalist repertoire
إسرائيل ورم سرطاني	Israel is a cancerous tumour
إسرائيل اغتصبت أرض فلسطين	Israel has raped the land of Palestine
ما أخذ بالقوة لا يسترد إلى بالقوة	What has been taken by force can be returned only by force
المجتمع الدولي منحاز لطرف إسرائيل	The international community is biased in Israel's favour
السلام مع الكيان الصهيوني الاستعماري مرفوض	Peace with the Imperialist Zionist entity is rejected
ضعف و عدم شرعية الأنظمة العربية	The weakness and illegitimacy of so called "Arab" regimes
المرجع الإسلامي	**The Islamist repertoire**
الصراع مع إسرائيل صراع عقائدي	The conflict with Israel is religious
الرجوع للإسلام الوسيلة الوحيدة لمواجهة إسرائيل	Returning to Islam is the only means to confront Israel
قيام إسرائيل مذكور في القرآن الكريم	Israel's ascendance is prophesied in the Quran
نكبة فلسطين نتيجة التآمر على نظام الخلافة	The Palestinian catastrophe is the result of the abandonment of the Caliphate
كما حرر المقاومة الإسلامية جنوب لبنان ستحرر فلسطين	In the same way that Hezbollah liberated south Lebanon Palestine will be freed
لا فائدة في المعاهدات مع إسرائيل – ف وسفهم الله في القرآن بأنهم إذا وعدوا اخلفوا وإذا أتمن خان	There is no point in negotiating with the Israelis – Allah has described them in the Quran as breakers of trust and promises
المرجع اللبرالي	**The Liberal repertoire**
دولة إسرائيل أصبحت حقيقة يجب التعامل معها	The State of Israel has become a reality we must deal with
اليهود لديهم حق تاريخي في العيش في أرض فلسطين التاريخية	The Jewish people have a historical right to live in the land of Palestine
لا يوجد بديل للمفاوضات مع إسرائيل	There is no alternative to negotiating with Israel
السلام مع إسرائيل خيار استراتيجي	Peace with Israel is a strategic choice
الحقد والشعارات وتشدد الأعمى لا يؤدي إلى شيء	Hatred, rhetoric and blind extremism has got us nowhere
مبروك لإسرائيل	Congratulations to Israel

language forums (see Table 2) would not be acceptable in the more plural space of English-language discussion; therefore, such comments received are likely to be removed and would not be available for analysis.

Repertoire analysis of the BBC Arabic discussion of the 60th anniversary of Al-Nakba

To tackle these problems of comparison we used two main kinds of analysis: of discourse/repertoire, and of interactivity. We had intended primarily to compare interactivity across the forums, but because the Arabic forums were much less interactive, we analysed the discourse of the largest Arabic forum in more detail to see what was going on. We found that most contributors tended to make statements using a rather limited cluster of set phrases, which we identified as drawing on three repertoires: the Islamist repertoire, the nationalist repertoire and the liberal repertoire.

Phrases and metaphors in the repertoires could be found throughout the discussion forum with regularity of form and composition. The key phrases in each repertoire are identified in Table 2.

Of the 829 contributions on the forum, 93 percent (or 775) participations correspond to one of the three repertoires identified.

As Table 3 shows, the overwhelming majority of participations corresponded to a contemporary version of the pan-Arab nationalist discourse (67 percent). Although liberal repertoires were used in 21 percent of participations, those using them received predominantly negative responses from other users. We shall say more about the predominance of these repertoires in the discussion of the comparison of English and Arabic forums below.

English-language forums

The English language forums considered reflected the fact that English language audiences were concerned with marking not just the anniversary of Al-Nakba, but also that of the state of Israel. Hence discussion included both Israeli and Palestinian issues and perspectives. The two BBC debates chosen were "What Does the Future Hold for Israel?" (*Have Your Say*; run by the BBC's News division, bbcnews.com) and "Do Palestinians Need to Accept There Can Never Be a Home-coming?" (BBC *World Have Your Say*, run by the World Service). The Al-Jazeera English forum from 15–16 May 2008 was not available for analysis, so a similar topic from later in the year, "Does Israel Need a Different Kind of Politics?", was chosen. The occasion for this debate was Israeli leader Tzipi Libni's call for early elections after her Kadima party's failure to form a coalition government on 26 October 2008.

A significant feature of these debates in terms of the role for the forums as a public sphere, where interlocutors unlikely to meet face to face are able to discourse, is that they provide a forum for some Arab–Israeli and Muslim–Jewish dialogue, with participants from Kuwait, Qatar, Iraq and Iran interacting with Israeli participants in each

Table 3. Proportion of comments to BBC Arabic debate, by repertoire type.

	Repertoires	
Islamist	Nationalist	Liberal
87	523	165
12%	67%	21%

case. However, these voices hardly dominated; US participation was highest except in the case of *WHYS*, which attracts large African audiences. The proportions of contributors by country of origin (where known, by self-report), are given in Table 4.

The two BBC debates were hosted by different parts of the corporation, although both were linked to World Service broadcast content. *HYS*, hosted by bbcnews.com, is linked to the BBC's main news site (and hence attracts the highest volume of traffic in our study), and is a fully pre-moderated message board. This means that all content is read before it is posted; the proportion of material posted varies from 10 percent to 90 percent. Although some responses to others' comments are published, pre-moderation limits the conversational aspect of the format.

These debates provide a good example of the differences in tone discussed above between *WHYS* and bbcnews.com, with *WHYS*'s reactive moderation policy allowing for a much freer flow of conversation, and our longitudinal observation show greater regular participation. The connection between blog (where the content of the day's debate is discussed), broadcast radio phone-in show and follow-up forum has arguably created some sense of online community, and greater trust and intimacy is arguably evident between participants. Consider the following exchange:

Steve (UK) May 15, 2008 at 4:16 pm

There are refugees from every other conflict in history, and they all moved on. There are ethnic Germans got kicked out of the Czech republic after World War 2. [...] Israel isn't going to accept a large group of hostile people within Israel proper, just like the Jews expelled from Muslim countries realise it wouldn't be the wisest move to go back to those countries, yet history ignores Jewish refugees because Israel took them in, and didn't keep them in refugee camps like the Arabs do with Palestinians.

[...]

Lubna (Iraq) May 15, 2008 at 5:14 pm and 6:56 pm

Hello Precious Steve ... you go back to Precious Ros' post "Happy Birthday Israel?" and read a comment of mine (no. 87 I think) in which I described how Baghdadi Jews used to live in Baghdad before 1948 ... The right of returning back home again for Palestinian refugees never dies no matter how much time has passed ... the Jews were nearly 2000 years put of their homeland, and they still wanted to return.

Table 4. Proportions of contributors by country of origin, English-language forums on Al-Nakba.

	WHYS (n = 44)	*HYS* (n = 346)	Al-Jazeera (n = 44)
United States & Canada	16%	17%	37%
Other sub-Saharan Africa	30% (of which Nigeria 16%)	8%	12%
Western Europe	7%	16% (of which UK 11%)	11%
Israel	2%	6%	7%
Middle East and North Africa (except Israel)	11%	9%	-
South Asia	-	8%	5%

This sense of online community influences the kind of interaction that the forum contains, and especially increases the proportion of participants who ask questions of others, and who show evidence of careful (if critical) reading of others' comments – for example, by quoting them. The Al Jazeera site achieved a sense of dialogue in a different way, by having many contributions from a small number of active individuals (who could be said to dominate the discussion); in this situation, there are no questions (other than rhetorical) of other participants, and no quoting, but a high proportion of replies.

Comparison of English and Arabic forums

We compared Arabic and English forums in terms of the forms of communication that they exhibited. Dahlberg (2001) had evaluated the contribution of online political discussion forums to democratic practice in terms of the extent to which contributions match the ideal speech conditions in Habermas's theory of communicative action (1987), seeking to measure features such as (i) "exchange and critique of [...] moral-practical validity claims"; (ii) "reflexivity"; (iii) "ideal role-taking"; (iv) "sincerity"; and (v) "discursive inclusion and equality" (Dahlberg 2001, 615–620). We used some similar measures, including the amount of interaction, the extent to which others' arguments are engaged with, and the number of contributions which challenge the framing of debates. We recognized the limits of such an approach exposed by critics of Habermas's public sphere model, as indicated above, because a broader view of politics suggests that the solidarity-creating role of forums – for example, generated by participants referencing a particular discourse in the contributions – and less easily quantifiable contributions such as the submission of video clips, also need consideration. However, such arguments do not invalidate measures of interaction – rather, this communicative action is a significant dimension of democratic will formation, but one that needs to be considered alongside others. Our results are shown in Figure 1.

Figure 1 shows that English-language forums show higher rates of interactivity between participants, measured by the proportion of comments that reply to other participants, where the forms of moderation are the same (*HYS* uses full moderation which reduces the proportion of replies). Other measures of interactivity include "incorporation", or showing reflective engagement with another comment, and "search", meaning asking questions of other contributors. Challenges to the way in which debates are framed by editors are shown in the "frame" results, which are also higher (though

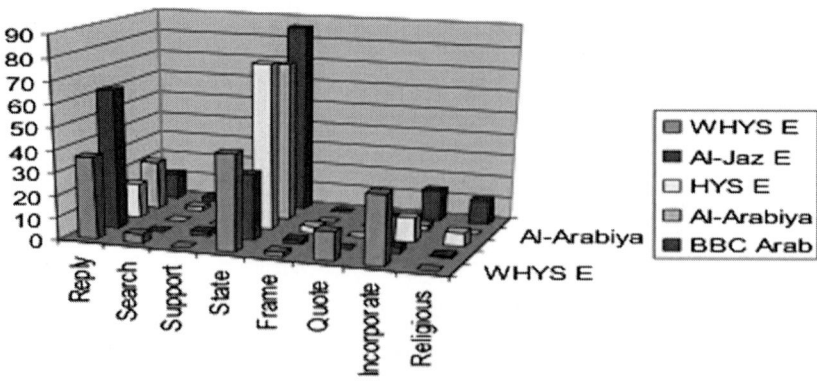

Figure 1.   English and Arabic Forums on Al-Nakba commemorations by contribution type.

still a small proportion) for English forums. This evidence suggests that the English-language debates are more interactive, and include more comments which show reflective engagement with other participants.

So, how do we explain these differences between interactivity on English and Arabic forums? One argument is that English, as a global lingua franca, is more likely to attract a wider variety of people to participate in debates. These people consequently give rise to a larger number of discourses and require a more complex negotiation with the diasporic other. As a result, those debating the Arab–Israeli conflict in English are likely to deploy language in a way that presents them as rational, compromising, empathetic but dispassionate, and perhaps above all as deploying a separation between politics and culture. In contrast, participation in the Arabic forums, at least on this issue, is more about showing solidarity. Contributions to a forum where a more limited range of political positions is likely to be displayed, and from contexts where forums for other forms of public participation are limited, take on different characteristics. More radically, one could argue that the distinction between politics and culture in Habermas's model of communicative action (from which our measures of interaction derived) disqualifies forms of expression that are traditionalist, mythico-magical or religious-metaphysical from being valid communicative acts – or, in our case, as not containing enough incorporation, seeking, seeding and replying. Ultimately, it could be argued, this approach rules out the validity of other "lifeworlds". Bakhtin's (1981) concern with the aesthetic, the axiomatic and the emotional, alluded to in the introduction to this article, could be used to develop measures of interaction which reflect these dimensions of communication, and in turn might help to validate and understand forms of expression that have not been addressed by academic analyses of the properties of political forums, and are also often considered a problem by policymakers in the west.

Conclusion

We have argued that the development of online forums is embedded in wider political, commercial and technological shifts that have created an elusive series of networks and alliances that shape the production and dissemination of news and information. The framing of debates, together with the pathways that connect and radiate out from them, creates the feeling of freedom of movement and access to information across a wide range of topics and regional areas. However, we have shown that these networks may rather be viewed as routes to a selection of officially sanctioned sites, and they may reflect the BBC and the FCO's desire to "steer" public attention away from more contentious sources of information. Such a desire may be argued to be deeply embedded in postcolonial relations, perhaps unsurprisingly most visible in the Middle East. We have shown in our comparison of English and Arabic forums that not just the BBCWS's framing but academic methods of comparison and evaluation also arguably reflect postcolonial relations, and particular attitudes to the political role of religion and culture.

On the other hand, our findings also point to social media as enabling some partial fulfilment of the utopian dreams of worldwide participation of some of the Web's pioneers, and alluded to in the BBCWS and FCO's rhetoric of "global conversation". While such rhetoric conceals power relations, these forums do function as a series of global public spherecules and screens in several if limited significant senses. First, they provide opportunities for participation in debate for those for whom such opportunities are otherwise restricted (e.g. political discussion in Egypt under the former Mubarak regime). Second, they bring into interaction individuals with common concerns who would otherwise be unlikely to converse (e.g. Arabs and Israelis). Third, in spite of constraints on interactivity imposed by technology, policy, and the news media discourses in which debates are often framed, many interactions illustrate the capacity of the format to enable complex argument, a crucial element of a discursive public sphere, even if argument is not the only reason for participation – solidarity is also important, as the Arabic forums showed. The forums also function as diasporic contact zones, in the sense that people of Arabic linguistic and cultural heritage from across the world are able to discuss with others in their region of origin.

Note

1. On 25 June 2012 *WHYS* posted its final blog, citing as the main reason for its closing down a shift to newer technologies: "the vast majority of our regular contributors have made it clear they'd prefer to take part in WHYS on our facebook and twitter pages". Discussions did continue on those pages (Atkins 2012).

References

Altheide, David L. 1991. "The Impact of Television News Formats on Social Policy." *Journal of Broadcasting and Electronic Media* 35 (1): 3–22.

Asad, Talal. 1999. "Religion, Nation-State and Secularism." In *Nation and Religion: Perspectives on Europe and Asia*, edited by Peter van der Veer and Hartmut Lechmann, 178–196. Princeton, NJ: Princeton University Press.

Atkins, Ros. 2012. "We're Not Here, We're There…" Accessed April 27, 2013. http://www.bbc.co.uk/blogs/worldhaveyoursay/

Bakhtin, M.M. 1981. *The Dialogical Imagination.* Translated by Caryl Emerson and Michael Holquist. Austin: University of Texas Press.

BBC Trust. n.d. "Purpose Plan for Delivering the BBC Public Purpose: Bringing the UK to the World and the World to the UK." Accessed August 8, 2013. http://www.bbc.co.uk/aboutthebbc/insidethebbc/whoweare/publicpurposes/world.html

Benahbib, Seyla. 1992. "Models of Public Space: Hannah Arendt, the Liberal Tradition, and Jürgen Habermas." In *Habermas and the Public Sphere*, edited by Craig Calhoun, 73–98. Cambridge, MA: MIT Press.

Benson, Rodney D. 2009. "What Makes News More Multiperspectival? A Field Analysis." *Poetics* 37 (5–6): 402–418.

Beyers, Hans. 2004. "Interactivity and Online Newspapers: A Case Study on Discussion Boards." *Convergence: The International Journal of Research into New Media Technologies* (10): 11–20. doi:10.1177/135485650401000403.

Bohman, James. 2004. "Expanding Dialogue: The Internet, the Public Sphere and Prospects for Transnational Democracy." *The Sociological Review* 52 (1): 131–155. doi: 10.1111/j.1467-954X.2004.00477.x.

Cottle, Simon. 2006. *Mediatized Conflicts*. Buckingham: Open University Press.

Dahlberg, Lincoln. 2001. "The Internet and Democratic Discourse: Exploring the Prospects of Online Deliberative Forums Extending the Public Sphere." *Information, Communication and Society* 4 (4): 615–633. doi: 10.1080/13691180110097030.

Dahlberg, Lincoln. 2005. "The Corporate Colonization of Online Attention and the Marginalization of Critical Communication?" *Journal of Communication Inquiry* 29 (2): 160–180. doi: 10.1177/0196859904272745.

FCO. 2006. "Active Diplomacy for a Changing World: The UK's International Priorities." White Paper presented to Parliament by the Secretary of State for Foreign & Commonwealth Affairs. Accessed October 1, 2007. http://www.fco.gov.uk/Files/kfile/fullintpriorities2006.pdf.

Fraser, N. 1992. "Rethinking the Public Sphere: A Contribution to the Critique of Actually Existing Democracy." In *Habermas and the Public Sphere* edited by C. Calhoun, 109–142. Cambridge, MA: MIT Press.

Gaus, Gerald, and Kevin Vallier. 2009. "The Roles of Religious Conviction in a Publicly Justified Polity: The Implications of Convergence, Asymmetry and Political Institutions." *Philosophy and Social Criticism* 35 (1–2): 51–76.

Gitlin, Todd. 1998. "Public Sphere or Public Spherecules." In *Media, Ritual and Identity*, edited by Tamar Liebes and J. Curran, 168–174. London: Routledge.

Habermas, Jürgen. 1987. *A Theory of Communicative Action*, Vol. 2. Cambridge: Polity.

Habermas, Jürgen. [1962] 1989. *The Structural Transformation of the Public Sphere*. Cambridge: Polity.

Habermas, Jürgen. 1996. *Between Facts and Norms*. Cambridge: Polity.

Habermas, Jürgen. 2008. "Notes on Post-Secular Society." *New Perspectives Quarterly* 25 (4): 17–29. doi: 10.1111/j.1540-5842.2008.01017.x.

Hain, Peter. 2000. "The BBC World Service: A Service for Freedom." Accessed October 1, 2007. http://www.fco.gov.uk/en/newsroom/latest-news/?view=Speech&id=2148677

Heller, P. 2009. "Democratic Deepening in India and South Africa." *Journal of Asian and African Studies* 44: 123–149.

Herbert, David. 2003. *Religion and Civil Society*. Aldershot: Ashgate.

Linaa Jensen, Jakob. 2003. "Public Spheres on the Internet: Anarchic or Government-Sponsored: A Comparison." *Scandinavian Political Studies* 26 (4): 349–374. doi: 10.1111/j.1467-9477.2003.00093.x.

Jones, Quentin, and Rafaeli Shaizaf. 2000. "Time to Split, Virtually: 'Discourse Architecture' and 'Community Building' Create Vibrant Virtual Publics." *Electronic Markets* 10 (4): 214–223. doi: 10.1080/101967800750050326.

De, Luca, Kevin, and Jennifer Peeples. 2002. "Public Sphere and Public Screen: Democracy, Activism and the 'Violence' of Seattle." *Critical Studies in Media Communication* 19 (2): 125–151.

Mayhew, Leon H. 1997. *The New Public: Professional Communication and the Means of Social Influence*. Cambridge: Cambridge University Press.

Milioni, Dimitra L. 2009. "Probing the Online Counterpublic Sphere: The Case of Indymedia Athens." *Media, Culture and Society* 31 (3): 409–431.

Mitra, Ananda. 2005. "Creating Immigrant Identities in Cybernetic Space: Examples from a Non-resident Indian Website." *Media, Culture and Society* 27 (3): 371–390.

Norris, Pippa, and Ronald Inglehart. 2004. *Sacred and Secular: Religion and Politics Worldwide*. Cambridge: Cambridge University Press.

Poole, Elizabeth. 2002. *Reporting Islam: Media Representations of British Muslims*. London: I.B. Tauris.

Rajagopal, Arvind. 2001. *Politics After Television: Hindu Nationalism and the Reshaping of the Public in India*. Cambridge: Cambridge University Press.

Sandell, Mark. 2008. Interview with *World Have Your* Say editor Mark Sandell, conducted and recorded by Tracey Black, 20 March.

van der Veer, Peter. 1999. "The Moral State. Religion, Nation and Empire in Victorian Britain and British India." In *Nation and Religion: Perspectives on Europe and Asia*, edited by Peter van der Veer and Hartmut Lehmann, 15–43. Princeton, NJ: Princeton University Press.

Wright, Scott, and John Street. 2007. "Democracy, Deliberation and Design: the Case of Online Discussion Forums." *New Media and Society* 9 (5): 849–869.

Panopticons within panopticons: surveillance inversions in Willie Doherty's video installations

Paula Blair

Queen's University, Belfast, Northern Ireland

From the understanding that postcolonialism is linked to power structures, and that surveillance activity is a means for knowledge acquisition, this article considers different ways of seeing/being seen as avenues for gaining control. It addresses such notions with reference to Northern Ireland, a region in the UK confused over its colonial/postcolonial identity, and takes into consideration its geopolitical position as well as its internal complex web of sociopolitical relations. It argues that control occurs at both state and popular culture levels, and that mass media involve a different kind of colonization – which can be seen as psychological. Advancing technologies allow for different kinds of invasion/occupation, and technomediated modes of modern living come complete with adapted exclusions and inclusions. These issues are explored through analysis of video installations by Derry artist Willie Doherty, and with reference to Belfast writer Ciaran Carson's essay "Intelligence" from *Belfast Confetti* – both Doherty and Carson focus on "watching" in the (conflicted/occupied) cityscape and the way it affects perception, language and identity. These issues of control and status are discussed here in the light of Foucault's theorizing of the panopticon and Bhabha's notions of mimicry and ambivalence.

Kevin D. Haggerty and Amber Gazso (2005, 183) point out in their study of terrorist threats and surveillance that the gaze of information gathering, regardless of its origin, causes psychological change in its objects through their awareness of being watched. It is this manner of looking, and its effects on the object, that are considered in this article, which examines several video installations by Derry-born artist Willie Doherty, namely *Blackspot* (1997), *Control Zone* (1999), *The Only Good One is a Dead One* (1993) and *Sometimes I Imagine It's My Turn* (1998). In dealing with issues of urban panopticism and the presence of social media, these installations indicate the changes in landscape, visual communication and language caused by the application of observational systems and mechanisms at state and (anti)social levels.

When the British government installed advanced security applications in Northern Ireland during the Troubles from the late 1960s, the region was transformed into a "laboratory of power" (Foucault 1977, 204). Over the course of three decades, this activity reshaped rural landscapes and urban architecture with watchtowers, barricades, prisons and high-rise accommodation. This notion of physical division filtered down through communities and domestic levels, meaning that everyday objects amounted to what Ciaran Carson refers to in "Intelligence" (an essay included in his poetry collection *Belfast Confetti*) as "panopticons within panopticons" (1989, 79) – a reference to Michel Foucault's appropriation of Jeremy Bentham's prison design based on a non-returnable gaze, where "[v]isibility is a trap" (Foucault 1977, 200). Changes occurred in people's habits of watching, due to growing access to information and communications media. Not only could local inhabitants suddenly see what was happening around the world almost in real time, but the rest of the world could witness Northern Irish events in the same manner. One function of the camera in current affairs and security is to gather information about the subjects of the image – who can thus be easily objectified.

As a 12-year-old, Doherty witnessed events during Bloody Sunday, 31 January 1972, when 13 civilians were killed and many injured after the British army opened fire during an outbreak of violence that coincided with an unauthorized civil rights march in his hometown of Derry City. Lingering memories of this witnessed trauma, and responses to news media's role in the conflict, permeate Doherty's oeuvre. He began in the 1980s with alternative approaches to landscape photography often overlaid with text, and progressed in the 1990s to projections and video installations. Twice nominated for the Turner Prize, he has attracted international acclaim with exhibitions and site-specific video-making throughout the UK, Ireland, Europe and North America, and represented Northern Ireland at the 2007 Venice Biennale, for which perhaps his most famous work, *Ghost Story* (2007a), was commissioned.

Belfast-born writer Ciaran Carson is the director of the Seamus Heaney Centre for Poetry at Queen's University, Belfast. He is an award-winning author and translator of poetry collections, epic poems, prose non-fiction and novels. As Temple Cone states, in the essay "Intelligence" to which this article refers, "Carson identifies the city of Belfast with the Utilitarian philosopher Jeremy Bentham's 'Panopticon,' a model reform prison composed of individual cells equally visible from a central tower" (2006, 68). This identification depicts the city as an arena in which power struggles for control between the security forces, Loyalist paramilitaries and the Provisional IRA continually take place. This creates what Carson describes as "the mixed mode – panopticons within panopticons" (1989, 79). However, Cone (2006, 68–69) points out that rather than remaining fixed and static like Bentham's model, the panopticon of Belfast is subject to constant reconstruction and remapping due to fluctuating power relations, the surveillance of surveillance systems, and the damage done to the city's infrastructure as a result of terrorist acts. A sense of "doubleness" emerges in Carson's writing, just as it does in Doherty's video installations, particularly in his descriptions of methods of obstructing or enabling vision and how the individual can shift between, or simultaneously assume, the roles of observer and observed. Moreover, both artists examine the behavioural effects of the self-policing structure of the panopticon – the fear of being watched is present whether the central tower is occupied or not.

Collectively, many of Doherty's installations criticize the desire for power and knowledge that gives rise to technological innovations. The motivation behind such advances is more significant than the need to define and control territory. As well as applying power/knowledge mechanisms to protect Britain's sovereignty in Northern Ireland, the experiments produced a potentially transferable model for social control. The use of Northern Ireland as a "laboratory" resulted in confusion over individual identities in an already ambiguous region – ambiguous in that it is held in stasis between colonial and postcolonial positions. During the conflict, not only was the population monitored by centralized state observation, but the state's observational techniques were often appropriated by "anonymous and temporary observers" (Foucault 1977, 202). In the context of the Troubles, these observers include non-state organizations such as paramilitary groups, who largely monitor their "own" communities from within, as well as keeping an eye to the "other" community and operatives of the state. Such imitation of the state's ways of watching leads to heightened confusion between reality and simulacra. By using Foucault's discussion of the panopticon and Homi Bhabha's discourse on mimicry to discuss Doherty's visual critiques of state observation, I aim to emphasize Northern Ireland's non-fit with constructs of Britishness and Irishness.

Northern/Irish identities and visual culture

Ireland is not unique in being a conflicted land or (former) British colony. Writing on Ireland in the 1990s, David Lloyd acknowledged the "fundamental dislocation" of "a culture which is geographically of Western Europe though marginal to it" (1993, 2). This same culture is "increasingly assimilated to that Europe, while in part still subject to a dissimulated colonialism" further complicated by continuing emigration. These factors combine to create what Lloyd terms as "the anomalous states of a population whose most typical experience may be that of occupying multiple locations, literally and figuratively" (1993, 2). Given that anyone born in Northern Ireland can claim *either* British and Irish nationalities or *both*, he or she experiences a similar "doubleness" without leaving the country.

While much attention is still paid to notions of Irish national identity, including its relation to the Irish diaspora, the more complex Northern Irish national identity is only recently becoming recognized, particularly since a desire for such an identity emerged in the 2011 census. In his introduction to the *Nationalism, Colonialism and Literature* Field Day pamphlet that was published when the Troubles showed no sign of ceasing, Seamus Deane (1990, 3–19) provides an example of a widespread lack of recognition that the Northern Irish identity is trapped in the fault line between contrasting British and Irish nationalisms. Discussions such as his, which in this instance concern the anxiety of individual identity within a community and its history, tend to omit the "between-ness" of the Northern Irish identity as an option or, worse, assume there is a straight split between the "British" and "Irish" populations in the region. The dichotomy between Britishness and Irishness that pervades Deane's essay shares further binaries between Protestantism and Catholicism, the colonizer and colonized, and, on a state level, the observer and observed. However, his essay supports the notion of mimicry that is discussed in this article with regard to Doherty's video installations. Deane states that "Irish nationalism is, in its foundational moments, a derivative of its British counterpart", and that its prevention "from being a movement towards liberation" lies in the fact that it is "a copy of that by which it felt itself to be oppressed" (1990, 7–8). I shall suggest below that Doherty's *Blackspot* and *Control Zone*, in which he assumes the responsibility of monitoring Derry neighbourhoods while continuing himself to be monitored by the state, explore just these complicated identities that derive from the see/being seen dyad. He questions state observation while mimicking its methods.

Homi Bhabha argues that "colonial mimicry is the desire for a reformed, recognizable Other, *as a subject of a difference that is almost the same, but not quite*", and therefore "the discourse of mimicry is constructed around an *ambivalence*" (1994, 86; emphasis in original). The effectiveness of mimicry relies on its display of difference. Its duality lies in "a complex strategy of reform, regulation and discipline, which 'appropriates' the other as it visualizes power" (Bhabha 1994, 86). This can be related to Deane's (1990) observation that Ireland's freedom is prevented because features of Irish nationalism reflect features of British nationalism; however, the power–knowledge relation in state observation is disrupted when aspects of the 'original' are appropriated by the 'mimic'. Bhabha goes on to describe how state visualizations of power that are countered with non/anti-state inversions of surveillance activities are indicative of disobedience. This disobedience echoes "the dominant strategic function of colonial power" that "intensifies surveillance" and threatens " 'normalized' knowledges and disciplinary powers" (Bhabha 1994, 86). Doherty's installations exploit these kinds of mimicry, as well as the residual signs of colonialism that remain visible in the architecture of security and control in landscapes and cityscapes in postcolonial Northern Ireland. Although most British army bases in Northern Ireland (apart from training barracks) have been decommissioned, many structures and observational mechanisms linger as spectral presences. CCTV's ubiquity iterates the idea that anyone can become a perpetrator or victim at any time – it watches with expectation. Attempts by the state to stifle or expose crime and terrorist or paramilitary activity may lead to suspicions filtering through neighbourhoods. In Northern Ireland, such a way of life became normal, with security checkpoints throughout city centres and an increase in broader surveillance activity in the 1970s.

Francis McKee (2007) indicates that conflict in Northern Ireland presented an opportunity for experimentation with devices of mass control, including CCTV in public spaces. This "suspension of normal democratic life enabled successive British governments to use the region as a testing ground for new technologies, an experiment in creating a contemporary controlled society" (2007, 19). News media attention grew at the same time. These kinds of information exchange, and the perceptions of others they generate, are significant themes in Willie Doherty's 1990s video installations. Drawing both forms of watching together, McKee describes the Troubles as follows:

> Throughout the 1970s and '80s, a persistent series of sectarian killings and bombing campaigns became routine reality for the population of Northern Ireland. Many countries had experienced guerrilla war before, most recently Vietnam. What changed in Ireland was the blanket presence of broadcast media allied to the installation of sweeping military surveillance. […] Repeated on a weekly basis, the images of that conflict became ingrained in the psyche – the roadside execution, the ruins of a building, the military checkpoint, barricades, the petrol bomb, the fortified barracks. (2007, 20)

Dualities emerged daily between the media images people encountered, and what they witnessed through the frames of their windows. The tension between what is seen naturally in real time and what is replayed from recordings made with apparatus capable of altering the original image (by enhancement, zooming, blurring, colour grading, etc.), raises issues concerning personal perception and the psychological effects the act of surveillance. McKee (2007, 20) asserts that those effects are heightened by an awareness of observation, that stems from its integration into physical landscapes in the form of watchtowers and listening stations.

Technologies of watching and modes of division

Technologies of seeing/being seen and physical modes of division, such as walls, buildings and chain-link fences, create derivative verbal and visual languages. For instance, graffiti affiliated to paramilitary groups denotes who is and is not welcome in an area (often accompanied with threats of violence), and alters the object which it adorns: for example, a home's gable wall is transformed into a territorial boundary, so that language and visual signifiers become further modes of division. Ciaran Carson's essay "Intelligence" highlights similar dualities in the existence of everyday objects during the Troubles, ranging from "bread vans, milk-carts, telegraph poles" to "news-stands, camera tripods, ladders" and people, all of which were treated as barriers to be removed (1989, 79). Their removal, and therefore the constant remapping of space, is facilitated by "the provision of all manner of lighting devices" (1989, 79). In a similar way to that described by Carson, Doherty's gallery spectators are confronted by different ways of regarding the same thing. For example, *Blackspot* oversees Derry City's Bogside area from an actual military vantage point. The vehicles, houses, post boxes, lamp posts and people that appear in the frame become objects that obstruct the camera's vision, yet each to an extent also provides a central point of outward vision within the housing estate while it is enclosed within the gaze of the elevated camera. To apply the same logic to *Control Zone*, which in a similar style overlooks the flow of traffic on the city's Craigavon Bridge, each car could be seen as a smaller panopticon within the larger panopticon of the bridge – itself a dual object of division and connection – all caught in the camera's scope from the panoptic Derry walls. Each object has a practical, quotidian function, and yet can be used to obstruct or enhance vision, while the transitory spaces of roads and bridges keep people separated, or bring them together.

Where Carson scrutinizes changes in his native Belfast, Doherty concentrates on his home town of Derry, County Londonderry, where advancements in observation technologies have come after centuries of developments in structures of fortification and defence. Derry's walls were constructed by the military in 1618 and are continually updated with state-of-the-art surveillance equipment. Mark Ward states that "Derry's walls were never breached and serve today as both a symbol of heroic resistance and of colonial domination" (2007, 27). In addition to their being historical markers of colonialism, Derry and Belfast were included in the growth of high-rise accommodation, designed to rehouse the working classes, throughout British urban centres in the 1960s and 1970s. In Northern Ireland, the double phenomenon of internal fortification and external social control was further aggravated by class divides appearing within already segregated Protestant and Catholic areas, which resulted in both class and denominational tensions (Kelly 1996, 60–63). Derry's hilly terrain particularly lends itself to such architectural fortification. The walls containing the city evoke Foucault's description of public surveillance:

> The enclosed, segmented space, observed at every point, in which the individuals are inserted in a fixed place, in which the slightest movements are supervised, in which all events are recorded, in which an uninterrupted work of writing links the centre and the periphery, in which power is exercised without division, according to a continuous hierarchical figure, in which each individual is constantly located, examined and distributed among the living beings, the sick and the dead – all this constitutes a compact model of the disciplinary mechanism. (1977, 197)

The hierarchical figure in the case of Northern Ireland is the British government, exercising power through the army sent to watch, control and discipline communities from high vantage points provided by the cities' architectures. The "compact model of the disciplinary mechanism" is relevant to the idea of Northern Ireland's use as a laboratory for the applications of that mechanism.

To expand on the idea that Northern Ireland was used as a testing ground for surveillance technologies, Lloyd points out that the "technique[s] of surveillance and data-harvesting" deployed there had already been applied "in other British counter-insurgency campaigns" in Asian and African colonies (2011, 13). Yet, he states, "[t]he peculiarity of Northern Ireland was that it represented a colony formally incorporated into the metropolis" (2011, 13). In his discussion of Carson's essay on "Intelligence", Matthew Brown (2010) observes that CCTV in Northern Ireland has operated within colonial surveillance, and in order for it to do so, communal and residential areas were reshaped throughout the 1980s and 1990s to facilitate the state's observation and information gathering. Of this restructuring, Brown observes:

> Walls and concrete barriers have divided cities into observable (read manageable) zones; urban planning was choreographed by sectarian division; public-use institutions [...] tended to more or less abide by the implicit politics of ethnic zoning, while public murals provided aesthetic force to the underlying tensions within urban development. The fierce politics of metropolitan visuality created cityscapes where citizens became hostages to this ethnic zoning. (2010, 57)

The connection between urban planning and readable/manageable zones is significant in Doherty's single-channel durational videos *Blackspot* and *Control Zone*, both filmed from British army vantage points on the Derry walls. The army's CCTV network fed the government's power–knowledge cycle and changed citizens' behaviour. Awareness of the gaze prompted paramilitaries to seek new means of evasion, often by arranging their own systems of information-gathering – not unlike those monitoring them. Such imitation shaped what Carson (himself drawing on Foucault's trope) describes as "panopticons within panopticons", and defined politicized territories through internal levels of social inclusion and exclusion (1989, 79).

The totality of the army's surveillance, as a network of many varied systems, amounts to Foucault's "machinery that assures dissymmetry, disequilibrium, difference" (1977, 202). Acts of surveillance signify the presence of an "other" that should be feared, and suggest that observed communities are imbalanced. Rather than an existing imbalance activating surveillance, Foucault suggests that it is the activation of surveillance that itself guarantees the imbalance. CCTV as a panoptic "application" of exercised power allows for a reduction in the numbers in those operating it, while increasing the population which is subject to it. This leads to possibilities of intervention and prevention of misdeed. Its strength lies in its hidden ubiquity, which causes it to never have to intervene, only watch (Foucault 1977, 206). However, using CCTV in this manner has its limitations. As outlined by Haggerty and Gazso, these include "interpretive ambiguity" and the hindrance to "real-time intervention" caused by "the sheer volume of information produced by existing surveillance systems" (2005, 181). The first issue is problematic, given that what could be construed as suspicious behaviour is not always obvious to users of the system. When this is combined with the volume of information to be processed, these problems can lead to a reliance "on stereotypical assumptions about how a 'criminal' or 'terrorist' looks or behaves" (Haggerty and Gazso 2005, 182). Haggerty and Gazso argue that the "risk indicators" which stem from these assumptions "often appear to be little more than racial/ethnic stereotypes" (2005, 182). The dissymmetry and difference that Foucault claims are assured by panopticism are caused by such "shortcuts" within the system "by singling out certain classes of individuals for greater scrutiny based on stereotypical attributes of group behaviour, they reproduce and reinforce biases against entire categories of people" (2005, 182). Not only this, but the subjects' increased awareness of monitoring makes it less effective as they alter their behaviour accordingly to mimic actions of compliance to shield their resistance.

Some four decades after initial installation of army CCTV in Northern Ireland, during which time its removal has been ongoing and the most intense observation has ceased, attention turns to effects on the observed. Abandoned posts remain as visible markers of the state's power over its subjects, but collective and individual psychological scars are not so easily identified. Although his treatment of the theme emerges in a discussion about the work of Indonesian author Pramoedya Ananta Toer, Pheng Cheah's assessment of "publicness and the spectral gaze of state surveillance" is relevant to Northern Ireland and to Doherty's installations (2003, 331). Cheah alludes to "a ghost (the nationalist movement) that haunts the state and holds it accountable because modern political authority is based on reason" (331). Moreover, the "colonial state's vulnerability to international public opinion is a more objective form of haunting" (331), as seen in American media interest in the civil rights episode of the Troubles. Because the observational tools used by the state are now available in different forms to the public, the transparency that develops is likened by Cheah to "the visibility of the glass house" (331), a barrier allowing two-way vision, in and out. As Cheah notes, education simultaneously enables and manipulates communication. Through a duality in "technomediation", public and state share a desire to watch/be watched, and become "attached to each other as phantom doubles" (331). Bhabha's instances of colonial imitation come from an "area between mimicry and mockery, where the reforming, civilizing mission is threatened by the displacing gaze of its disciplinary double" (Bhabha 1994, 86). He asserts that the ambivalence of mimicry results in the "partial", "incomplete" and "virtual" presence of the colonial subject (1994, 86). Not only are they *the same, but not quite* (1994, 86), they are there but not quite there, just like the operators or users of Doherty's spectral camera.

Foucault (1977, 202) argues that within social panopticons, panoptic mechanisms can be applied by anyone, and that more "anonymous and temporary observers" are useful in the quest to create surprise and anxiety, and to gain more power over the observed. Attaining vision necessitates the removal of barriers designed to obscure and divide. The openness provided by alteration of structures (e.g. replacing solid walls with chain-link fences) exposes those under suspicion of wrongdoing, while solid barriers are positioned to "shelter peaceable passengers" (or perhaps sedated compliers). Vision is linked with exclusion of the other, while cover protects and includes those under suspicion or scrutiny. When that other happens to be the "native" within a colonized state, public and state levels of scrutiny generate different connotations. They require different implements that allow vision while creating barriers – for example, one-way mirrors, television screens, cameras, transparent riot shields/visors and infrared goggles. Where most tools provide mechanized sight, advancements in vision are not restricted to mechanical devices. As Carson observes:

> Everyone is watching someone, everyone wants to know what's coming next, so the lightweight, transparent shield was a vast improvement over the earlier metal one because visibility was greatly increased and – an extra bonus – gave better protection against petrol and acid bombs which could flow through the grill mesh of the metal type … . (Carson 1989, 78)

This adaptation of riot shields addresses the desire to view coupled with the necessity for protection, to see out but keep in. Similarly, advancements in digital technology that improve CCTV cameras work towards the same goal: preservation of the state through a one-way gaze.

To reiterate, surveillance activity during the Troubles was conducted by both the state and paramilitaries to undermine the "enemy", and in the interests of self-preservation. To an extent, this is true of media activity. Paramilitaries found ways of inverting state surveillance by exposing hidden watchers or blocking their gaze. Carson notes the head-height whitewashing of walls "so that patrolling soldiers at night are silhouetted clearly for snipers" and "that paint bombs are usually reserved for throwing at the vision blocks of APCs and armoured cars" (1989, 78). Progressing from the use of paint to aid or obscure vision, Carson describes how technology grants sight beyond the human eye, stating that "passive observation is possible even on the darkest of nights, since the ambient light is amplified by this Telescope Starlight II LIEI 'Twiggy' Night Observation Device" (78). The scenes made viewable by such a device are realized by Doherty's *Blackspot* and *Control Zone*, aided by light provided by the city.

Vision and identity within and without the cityscape

In *Blackspot* and *Control Zone*, the communities involved are as unaware of Doherty's recordings as they are of real CCTV activity. As a member of the Derry community, he assumes the responsibility of monitoring its neighbourhoods while continuing to be monitored himself by the state, which complicates the see/being seen dyad. He questions the amount and quality of information that can be gleaned from the state's lofty observation; by mimicking its methods, he exposes its banality. Filmed during sunset, *Blackspot* suggests that covert filming is ineffective without adequate light and that everyday neighbourhood objects become obstructions. The installation loop reiterates this point by condensing the cycle of light and darkness into a 30-minute time frame. The scene is tedious – so much so that the viewer questions the purpose of their watching. A camera's presence over an area indicates that observation is needed, but its presence is questionable when it is placed at such a distance from the scene, as is evident in the long-range zoom. Their association with the military prompts Matthias Mühling (2007) to compare the cameras on these walls to non-lethal weapons. He believes their presence generates suspicion and conveys power over those caught in their scope (30). With this gaze displaced into a gallery setting, the viewer is confronted with a mirror image of sorts. In the darkness, the dimly lit streets could be part of any residential area, including the viewers' own. Given that the neighbourhood is viewed through the subjective lens of the "military" camera, the viewpoint implicates the spectator within and without the gaze. *Blackspot* is intended for direct wall projection in a dark space (the spectator becomes sightless to the point of disorientation upon entering most Doherty installation environments). As darkness falls, the scene and space are punctured by artificial lights switching on in streets, cars and homes. Even in a large projection, the darkness and grainy texture render actions difficult to discern. Human figures are vaguely seen crossing roads, getting into cars or conversing in streets. The dark silence is a desensitizing experience that shifts between tedium and interest, depending on how carefully a spectator decides to watch; the installation's effectiveness relies on the appeal of the camera's watchfulness.

Control Zone exudes the same harmless nature in that nothing happens, with life continuing as normal. It is another static real-time 30-minute shot, this time overseeing the Craigavon Bridge on Derry's River Foyle from a distant vantage point. The river roughly separates the mainly Protestant/Loyalist community on the east bank, and

Catholic/Nationalist community on the west bank, accessible to each other, and divided, by the bridge. The strong telephoto lens used in filming makes vehicles crossing the bridge appear to drift vertically up and down on-screen, distorting the reality of what is viewed. The omission of sound means that, unlike other work by Doherty, there is no language to imply context. *Blackspot* and *Control Zone* seem innocent and uneventful, but, as surface images with disrupted or distorted vision, perhaps they cloak something more sinister. Intensive watching of 30 minutes of real-time footage might prompt viewers to question the content of street discussions in *Blackspot*, or where people are driving to and from in *Control Zone*, before considering geographical context.

During protests, Bogside residents barricaded themselves in and kept the RUC (Royal Ulster Constabulary), B-Specials and army out with makeshift barriers guarded by men of the community adopting a protective kind of self-surveillance by conversing only with those they knew well. In the case of Derry's walls, structures alone do not denote who is included or excluded within or without them; rather, the language applied to them does. Murals and certain words or acronyms spray-painted on walls signify territory, affiliation and exclusion. Text or speech accompanying any image will sway what is perceived from the visual element. Like the state itself, news broadcasts withhold as much as they inform. Doherty counterpoints text and image in various ways, using riddles and abstract terms in the monologues and captions, or replacing verbal language with the visual languages of moving image media.

Sometimes I Imagine It's My Turn depicts a body lying in a non-urban setting, yet a connection to urban life is implied through media fragments intercut with the camera's repeated discovery of the body. The video shows slow pans across a desolate forest floor, encountering a man lying face down on sodden, exposed earth. Some shots are handheld and grainy (apparently taken with an older video camera), while others are smooth and sharp (shot with a newer digital camera on a dolly). The cameras repeatedly come to and pass around the man's body, edging closer each time, but never reaching his head – the identity remains concealed. Inserts of blurred CCTV footage and black-and-white news coverage, perhaps relating to the missing person, disrupt the scene as helicopter rotors "chainsaw overhead" (Carson 1989, 78) – a common sound in Northern Ireland, evoking searching, deterrence and control. Typically, we rely on televisual, newspaper, radio and online news features to inform our daily lives and broader concerns, but this information is gathered by the media's scrutinizing eye – an eye that can fall on any one of us at any time. The man's identity and the reason for his being in the woodland are open to speculation. In the Troubles context he could be a paramilitary, an informant, an innocent caught in the wrong place at the wrong time, perhaps even a police officer or soldier whose body has been "disappeared". Independently of the associations given by the CCTV inserts and helicopter rotors, this man could equally have been a walker who suffered an accident or affliction, but the viewer jumps to conclusions from the text of edited fragments. The gallery spectator succumbs to the perceptions we are all unconsciously trained to generate due to the way the media transmits the information it has gathered about us.

Doherty's *The Only Good One is a Dead One* indicates possible effects on receivers of transmitted mediated information. It features a looped monologue describing imaginary scenarios as an involuntary response to observational activity, although it is unclear whether it is paramilitary or not. The young man speaking believes he is being stalked by a killer, and the paranoia he develops out of irrational fear causes him to "mimic" the stalker. His frequent allusions to the media indicate the extent to which information transmissions have become incorporated into his mindset. He believes what they tell him and expresses himself in language of mediation:

> As my assassin jumps out in front of me everything starts to happen in slow motion. I can see him raise his gun and I can't do a thing. I see the same scene shot from different angles. I see a sequence of fast edits as the car swerves to avoid him and he starts shooting. (Doherty 2007b, 61)

The man's subjective, point-of-view descriptions read as if they were scripted directions. They are spoken without inflection or emotion but use aggressive language and at first appear to contradict the uneventful images between which the spectator is positioned in the gallery. Viewers receive a mental storyboard of sequences overlaying the two actual onscreen videos filmed from a car's interior under cover of darkness, one parked on a street (held in an urban setting), the other driving on country roads (the mobility transcending politicized territories). The scenarios described in the voice-over switch from what the man imagines are his "assassin's" actions towards him, to fantasies of how he will "get him first":

> In the early morning the roads are really quiet … You can drive for ages without passing another car. […] It might be just as easy on the street. […] It should be an easy job with a car waiting at the end of the street … I've seen it so many times I could write the script. (Doherty 2007b, 61)

Doherty is renowned for ambiguity, and only the disembodied voice's accent evokes Northern Ireland. This, combined with the "us and them"/"he and I" mentality and the victim/perpetrator dyad woven throughout the speech, implies the Troubles.

As with the previously discussed works, there is no obvious sign of a camera operator in either screen of *The Only Good One*. The detached gaze leaves the viewers unsure as to which subjective position they find themselves in – that of the watcher, or of the watched looking out for the watcher. The man's imagined sights, sounds, sensations and feelings are recycled back into his paranoia. His fear comes from familiarity, as he refers to television coverage of tragic events and claims that his adversary reminds him of someone, perhaps a former classmate. He believes his "killer" monitors him by phone-tapping as well as following him, and justifies this by claiming that his friends hear strange noises when calling him. He thinks this person rings to check whether he is home, yet claims to have "walked right past him" without being noticed. In tension between the stalker/stalked identities, the man feels he knows his follower and says, "I'm sick of looking at him", yet the "killer" seems not to know him, and it becomes apparent that his paranoia is fuelled by hearsay and media stories. The Troubles are not alluded to, but violence, opposition and loss are prominent, as is a sense that they are nearby and familiar, though not quite tangible: "I see these horrific events unfold like a scene from a movie … ". The way he likens news coverage and violent fantasy to cinematic sequences indicates his inability to distinguish reality from fiction. To him, violence becomes spectacle; it is exciting rather than frightening, and so familiar to him that he could "write the script". Instead of being desensitized by frequent violence on news broadcasts, television programmes and films, this man fancies himself a target and admits, "I've had some really irrational panic attacks … There is no reason for this but I think that I'm a victim." He places himself in others' positions when recalling a news story of "a particularly savage and random murder" in Belfast and "can't stop thinking about the awful fear and terror [the victim] must have felt" (Doherty 2007b, 60–61). This young man succumbs to a murder fantasy fetish; extreme levels of watching have brought about a schizophrenic paranoia and a split personality.

Identity crises and the gallery as site of intervention

In *The Only Good One*, Doherty confuses the spectator's sense of identity as much as the young man's, by inviting identification with both perpetrator and victim. When viewing such work, spectators become the surveyor/watcher/voyeur, but upon emerging from the gallery they resume their surveyed/watched/observed identity. Doherty urges spectators to consider ambiguous multiplicities of character from a psychological standpoint. It is with such evocation that the gallery potentially becomes an interstitial space, in which the position of the other is considered, accepted and understood through encounter. Viewers of any artwork are naturally drawn towards human presence or actions, only traces of which are often discernible in Doherty's work. In the pieces under discussion, the artist – to whose work text and speech are usually integral – chooses diegetic sound that is almost silence, although the titles present at once specific and ambiguous connotations. As Jean Fisher states, Doherty's work is frequently "devoid of the human figure and yet inscribed with the anxiety of an unseen observer" (2006, 35). In the films there are traces and fragments, but an absence of bodies – no holder of the gaze that the audience is aware of, simply a spectral sense of one. Foucault claims that "[t]he Panopticon is a machine for dissociating the see/being seen dyad: in the peripheral ring, one is totally seen, without ever seeing; in the central tower, one sees everything without ever being seen" (1977, 201–02). Fisher characterizes this dyad in Doherty's work as "a question of both position and disposition: *I see you in the place I am not*" (1990, n.p.; emphasis in original). In *Blackspot* and *Control Zone* little is seen, suggesting that the act of watching in this context is impotent. In the gallery installation, spectators find themselves as holders of the gaze. It is a gaze they do not choose, but hold nonetheless.

Doherty's works show identity on either side of state observation to be complex and interchangeable. Sociological theorist Gavin Smith (2009, 125–26) suggests that in the surveillance society, where the watched oscillate between opposite perspectives, the watchers circle from empowerment to disempowerment then re-empowerment in the information exchange. Just as Doherty assumes the centre's non-intervening gaze over Derry in *Blackspot* and *Control Zone*, the young man in *The Only Good One* assumes his "killer's" gaze and appropriates the power and control resultant from watching. Yet the "killer's" gaze is not grounded in reality and is as "sightless" as that of the CCTV camera. The power gained is unstable, owing to the young man's keenness to act upon speculation from what he claims to have observed around him. The issues arising from surveillance activity are internalized, and as such implicate the gallery spectator. By denying passive viewing, Doherty forces the gallery visitor to confront ambiguous identities, even regularly positioning them amid the two images in his double-channel works. When awareness of this coerced participation – another level of control – occurs, the gallery becomes an interstice, a site of intervention where the artist demonstrates the ability of "anonymous and temporary observers" to convincingly mimic state control.

Conclusion

Although only traces of the British army's intensive surveillance remain today, Northern Ireland may still be regarded as a surveillance testing ground, with ongoing research taking place at the Centre for Secure Information Technologies based in Queen's University's Institute of Electronics, Communications and Information Technology. The installations discussed here show that while the watched can become watchers, the correlating positions of the occupied and occupiers do not trade places in kind. The centre's gaze can be emulated, referenced, criticized and mocked by the periphery, but it remains the centre's gaze, upheld by a system of signs and submission to the status quo. While for many, surveillance provides greater security, and the desire to be visible leads to safety (Haggerty and Gazso 2005, 183), the works by Carson, Doherty and Foucault discussed in this article show that visibility is a potentially fatal trap which at the very least perpetuates a sense of "otherness" towards its subjects. Through artistic acts of "nonviolent resistance" (Cone 2006, 83) in works which interrogate the methods for navigating and monitoring movement in panoptic urban spaces, Carson and Doherty expose the ambivalent identities and slippery language that stem from attempts to mimic the state's techniques of observation in the struggle for power and control. There is no indication that observation will diminish; instead, it will escalate with advancing and merging communications technologies. New technologies allow for a different kind of invasion/occupation and the global village now comes complete with figurative twitching curtains. How can individuals discern personal identities when connected to a region such as Northern Ireland with its countless divisions – a region in a world where geographical, political, social and economic borders create disparate maps? The challenge this article poses to postcolonial studies is to address the different Northern Irish "trouble" of determining its place in that world.

Acknowledgements

The author would like to thank Des O'Rawe, Gary Rhodes, Martin McLoone, Caroline Magennis, Ole Birk Laursen, Florian Stadtler, Brian Rock and Willie Doherty for their guidance, support and insight.

References

Bhabha, Homi. 1994. *The Location of Culture*. London: Routledge.
Brown, Matthew. 2010. "Cities under Watch: Urban Northern Ireland in Film." *Éire-Ireland* 45 (1): 56–88.
Carson, Ciaran. 1989. *Belfast Confetti*. Oldcastle: Gallery Books.
Cheah, Pheng. 2003. *Spectral Nationality: Passages of Freedom from Kant to Postcolonial Literatures of Liberation*. New York, NY: Columbia University Press.
Cone, Temple. 2006. "Knowing the Street Map by Foot: Ciaran Carson's Belfast Confetti." *New Hibernia Review* 10 (3): 68–86.
Deane, Seamus. 1990. "Introduction." In *Nationalism, Colonialism and Literature*, edited by Terry Eagleton, Fredric Jameson, and Edward W. Said, 3–19. Minneapolis, MN: University of Minnesota.

Doherty, Willie. 1993. *The Only Good One is a Dead One*. Double-channel Video Installation. Dublin: Irish Museum of Modern Art.

Doherty, Willie. 1997. *Blackspot*. Single-channel Video Installation. Vancouver Art Gallery.

Doherty, Willie. 1998. *Sometimes I Imagine It's My Turn. Double-Channel Video Installation*. Dublin: Irish Museum of Modern Art.

Doherty, Willie. 1999. *Control Zone*. Pamplona: Single-channel Video Installation.

Doherty, Willie. 2007a. *Ghost Story*. Single-channel Video Installation. Belfast: Ulster Museum.

Doherty, Willie. 2007b. "Videography." In *Willie Doherty: Anthology of Time-Based Works*, edited by Yilmaz Dziewior and Matthias Mühling, 43–165. Hamburg: Hatje Cantz.

Fisher, Jean. 1990. "Seeing beyond the Pale: the Photographic Works of Willie Doherty." In *Willie Doherty: Unknown Depths* (unpaginated), edited by Christopher Coppock. Bristol: Taylor Brothers.

Fisher, Jean, Priamo Lozada, and Willie Doherty. 2006. *Willie Doherty: Out of Position*. Mexico City: Turner/A&R Press/ColecciÚn Jumex.

Foucault, Michel. 1977. *Discipline and Punish: The Birth of the Prison*. Translated by Alan Sheridan. London: Penguin Books.

Haggerty, Kevin D., and Amber Gazso. 2005. "Seeing Beyond the Ruins: Surveillance as a Response to Terrorist Threats." *Canadian Journal of Sociology* 30 (2): 169–187.

Kelly, Liam. 1996. *Thinking Long: Contemporary Art in the North of Ireland*. Kinsale: Gandon Editions.

Lloyd, David. 1993. *Anomalous States: Irish Writing and the Post-Colonial Moment*. Dublin: Lilliput Press.

Lloyd, David. 2011. *Irish Culture and Colonial Modernity 1800–2000*. Cambridge: Cambridge University Press.

McKee, Francis. 2007. "Smithereens." In *Willie Doherty: Anthology of Time-Based Works*, edited by Yilmaz Dziewior and Matthias Mühling, 19–24. Hamburg: Hatje Cantz.

Mühling, Matthias. 2007. "At the End of the Day it's a New Beginning." In *Willie Doherty: Anthology of Time-Based Works*, edited by Yilmaz Dziewior and Matthias Mühling, 30–33. Hamburg: Hatje Cantz.

Smith, Gavin. 2009. "Empowered Watchers or Disempowered Workers? The Ambiguities of Power Within Technologies of Security". In *Technologies of InSecurity: The Surveillance of Everyday Life*, edited by Katja Franko Aas, Helene Oppen Gundhus and Heidi Mork Lomell, 125–146. London: Routledge-Cavendish.

Ward, Mark. 2007. "In Place of Passing." In *Place of Passing*, edited by Julie Bacon, 20–48. Belfast: Bbeyond and Interface.

The borders of virtual space: new information technologies and European Islamic youth culture

Maruta Herding

German Youth Institute, Halle, Germany

The 2000s have seen the emergence of an unusual combination of western pop culture with an Islamic theme. Religious hip hop, street-wear with Islamic slogans or media for "cool" religious youths have become meaningful forms of expression for young practising Muslims in Europe. This article draws on the results of a wider research project on Islamic youth culture, but focuses on the role of new information technologies in this context. The empirical basis of this research is qualitative data collected in Germany, France and Britain in 2008–09. This comprises a collection of Islamic "virtual artefacts" ranging from, among other things, a French comedy web-site, a German youth platform and a British fashion blog, to in-depth interviews with the producers of these online spaces. The article analyses to what purpose the producers chose their respective means of expression. It also looks at how they use and shape existing forms of new information technologies by adding an Islamic feature to them. Arguably, the use of new information technologies greatly facilitates the growing movement of a European Islamic youth culture. But despite having obvious transnational potential, evidence shows that there is little interaction across national borders, let alone a significant awareness of similar trends in the neighbouring countries. The presence of young Muslims in virtual spaces is very often and deliberately a confined national project, shaped by their disputes with dominant society in the national context of their home country in the west.

Introduction

For young, religious Muslims growing up in Europe, issues of religion, identity and relations with dominant societies are increasingly significant. Until recently, these issues have often been addressed by others – for instance, elder generations or non-Muslims. Increasingly, however, young Muslim Europeans use new information technologies, such as comedy clips, fashion blogs and online discussion platforms, to negotiate issues of concern to their lives, such as cultural identity, and their participation in society. Such new technologies are not just entertainment, but also generate cultural products that are a deliberation of pressing social and political concerns affecting wider society and its relationship with Muslim minorities.

This article is based on sociological research on the development of Islamic youth culture in France, Britain and Germany (Herding 2013). That ethnographic study focused on the fashionable ways in which young European Muslims expressed their faith, and examined the manifestations of Islamic youth culture, the cultural references it put into practice, the reasons why people developed Islamic youth culture and the biographical, (sub)cultural, religious or societal reasons for this engagement.[1] In this article, I will focus especially on new information technologies within Islamic youth culture and on the producers' interview accounts, to provide new insights into the process of, and rationale for, their online production. Furthermore, the article will reflect on the context in which this trend occurs and thereby particularly focus on the relationship between global and local frames of reference.

In this field, the limited existing scholarship offers only a tentative overview of the scene of Islamic youth culture. For example, in her short introduction to western "cool Islam", Amel Boubekeur (2005) argues that a comfortable consumerism has replaced political Islamism, which allows Muslim pride to be shown in inconspicuous ways. However, her argument that this new culture is a form of secularization is not entirely convincing. She argues that because this new culture is based on elements from the secular sphere, it facilitates integration into western societies (2005, 12). Her conclusion may be true, and there is a matter of overlapping spheres at work, but this does not entail a secularized culture. By contrast, I argue that non-religious objects become imbued with religious meaning. Moreover, she narrows the movement down to an individualistic consumer culture and thereby ignores the entangled networks among youths, as well as the producers and the manifold intentions behind it.

Julia Gerlach was one of the first authors to offer a comprehensive overview of the movement in Germany that she first termed "Pop Islam". Her journalistic account gives intriguing empirical insights into a phenomenon that she claims may have great potential for integration due to a "remix of life-styles", the embrace of Islam and globalized popular culture (Gerlach 2006, 11). She points out that so-called "Pop Muslims" show a strong social commitment, which prompts her to assume that an engagement in Islamic pop culture and communal work will deter them from Islamist violence. But she also observes an explicit conservatism in occasional displays of anti-western opinions. As for the trend's origins, Gerlach assumes them to be rooted in the movement around Arab televangelist Amr Khaled, as Arabic-speaking German youths consume his sermons via satellite television and the Internet. Khaled's ideas and style, she claims, resonate with young Muslims in Europe because of the geopolitical situation that fosters anti-Islamic sentiments, coupled with their search for moral values in the Quran, as they embrace a young, fashionable outlook on Islam. By contrast, I will argue that the movement consists not only of a simple transfer of styles and ideas from the Arab world, but is deeply connected with the situation of young Muslims in Germany.

In the British context, Mandaville (2009) has provided an important introduction to this topic in an article on popular culture and Muslim youth. He relates the phenomenon of an emerging Islamic youth culture to issues of identity formation and sociopolitical attitudes, and detects several themes that dominate young British Muslims' popular culture. One such theme is having an Islamic version of mainstream popular culture, leisure activities and also middle-class lifestyles, partly as an alternative, but also in order to continue taking part in the mainstream. Another theme is the use of youthful expressions such as skilfully designed websites and flyers by organizations such as Young Muslims UK, but also by traditional religious scholars in order to reach out to young people (Mandaville 2009, 165–167). Mandaville correctly notes the urge of this age group to adapt Islam to the time and place in which they live, but he approaches it from the perspective of consumers only, without taking into account that of the producers to find out why they produce Islamic popular culture.

Thus far, existing scholarship has not focused particularly on new information technologies in the context of Islamic youth cultures. While new information technologies certainly facilitate the delocalization and global flow of cultures in this particular field, I shall argue that the use of such technologies is strongly embedded in local contexts. The potential global connectivity of new information technologies with an Islamic message is of less importance to the users and producers than the local ideas, actions or community that these technologies support.

New information technologies by and for young Muslims

The development of new information technologies has facilitated the emergence of a virtual space for young Muslims. These productions are marked by religious, Islamic messages and addressed to practising Muslim youths. These products have flourished, and continue to expand, in countries where Muslims are a minority, and particularly in European countries with the largest minority groups of Islamic background – France, Britain and Germany. Several examples of new information technology products from these countries will offer some invaluable insights into their use and the way in which they address issues of faith and youth.

Waymo (www.waymo.de) is an important new Internet platform for young Muslims in Germany. It offers similar features to YouTube and Facebook, allowing registered users to upload videos, pictures and audio files that have a relevance to them in a wider Islamic sense. Users can comment on the uploads and contact each other. The primary language is German, but some content is in English and, less often, in Turkish or Arabic, in which case a translation into German or English is usually provided. The fact that the community the platform addresses is at the same time German and Islamic is illustrated by the virtual "poke" function, similar to that found in other social networks such as Facebook. This way of greeting someone is called "salamen", a hybrid neologism consisting of the Islamic greeting *salam* and the suffix "-en" of German verbs. Waymo is closely connected to other representatives of German Islam – for instance, the Zentralrat der Muslime (Central Council of Muslims) and their website www.islam. de; the youth organization Muslimische Jugend in Deutschland (MJD); and the fashion company Styleislam, with which Waymo organized a Muslim comedy contest in 2008.

Waymo is a Web 2.0 feature that is developed further by the user – the creators of the website provide a platform, but content is added by consumers. Embedded videos and other content are not necessarily made by the users themselves, but come from other websites such as YouTube. This also means that users can post virtual material on a wide variety of subjects, which raises questions of moderation and control. The creators reserve the right to delete uploads and add a disclaimer on every page, where Waymo is not responsible for uploaded content or comments that do not reflect the opinion of Waymo.

The following case may be illustrative of why this might be necessary: at a random visit in 2010, the website featured an uploaded video of a radical religious leader in Britain, who claimed to disprove the belief that Jesus was the Son of God and thus tried to invalidate one of the central beliefs of Christianity. Another video of a Salafist lecture advocated strong gender segregation among a group of young male followers of the radical group Dawa Ffm. The radical preacher Pierre Vogel is also present in many videos, which might suggest a growing popularity of Salafist thought among some younger Muslims. Such videos are usually deleted if they do not comply with Waymo's rules of ethics, which specifically rule out extremist and political materials that could lead to misunderstanding and offence (www.waymo.de/rules). For the organizers it is difficult to determine when to delete such posts, especially when positive comments have been added. One such example was a video presentation on "beauty tips for sisters", which recommended that women should lower their gaze or adopt an attitude of modesty and wear less make-up – reinforcing strict and conservative gender roles. Yet this type of content is often posted by female users and is well received. Other uploads consist mainly of *nashids* (religious chant); pictures of mosques; videos of the hugely popular rapper, Ammar114; social awareness appeals (for instance, by Islamic Relief); comedy series, such as a Simpsons episode, if they contain Islamic references; and enthusiastic conversion accounts. This particular use of new information technology is particularly interesting for the study of young Muslims, because the final, yet constantly changing, product is interactively composed on a provided framework and of user-generated contents.

The French website A part ça tout va bien (www.apartcatoutvabien.com) is an online project promoting Islamic comedy. The name, meaning "apart from that, everything is fine", suggests a source of irritation that hinders Muslims from feeling fully at ease with dominant French society. Indeed, the project statement cites continuing conflicts between Muslims and non-Muslims – such as the ban on headscarves in schools, or practices such as divorce for the reason of non-virginity – as the reason for strong tensions dividing French public opinion. The project therefore tries to tackle mutual misunderstanding and distrust in a humorous way by encouraging people to laugh together, at each other and especially at themselves: "Qui a dit que les musulmans n'avaient pas d'humour?" (Who said Muslims don't have a sense of humour?) is the motto of this allegedly first website of Islamic comedy.

A part ça tout va bien is uses webisodes,[2] short videos which are exclusively available online and receive up to two million views per clip. The blog section allows for comments on the sketches, where users often offer enthusiastic feedback, with the occasional criticism that an episode was "un-Islamic" or misunderstands the basics of the French Republic. Among those mocked in the sketches are Islamic radicals, French racists, the French state for its unease with Islam, secular parents' astonishment at their children's return to religion, Muslims in France and in the Maghreb. A series produced by the project, "Islam School Welkoum" (www.islamschoolwelkoum.com), features three young French Muslims who attend a religious school in Morocco in search of "true" Islam, escaping the distraction, temptation and corrupted Islam they have come to know in France. Punchlines are directed at the hypocrisy of traditional religious education; thus in the course of the episodes the three wish for their former lives in France, retrospectively cherishing the individual freedom they used to enjoy.

None of the parties portrayed in the sketches escape caricature, which helps to modulate the tension between religion and secularism, Muslim and non-Muslim identities. The idea is that the tensions are often only aggravated by discussion, and while the sketches do not necessarily offer a solution as such, they foster an ironic and self-critical perspective on the situation in an attempt to depoliticize contentious issues between Muslims and non-Muslims, however fraught this process may be.

In the British context, the participants of the "Muslim Café", an online television programme run by Gazelle Media, also explored the relationship between entertainment and Islam (www.muslimcafe.tv; www.gazellemedia.co.uk). "Muslim Café" broadcasted debates from a Moroccan-style London café, where the participants discussed wide-ranging issues such as whether hip hop or Harry Potter were halal, their thoughts on dating and Islam, whether or not Muslims could have fun, and on the question "Can Muslim women have it all?" All topics revolved around religion and entertainment and arose from individuals practising Islam living in a non-Islamic environment. The discussants included Muslims from a variety of professional backgrounds and were mainly young people. Occasionally, non-Muslims also joined the debate. The programme ended in 2009.

Gazelle Media works for clients as diverse as the BBC, the British Council, the Young Muslims Advisory Group and community initiatives, and offers digitally produced content for television, radio, Internet and print. The company claims to provide "in-depth knowledge of multicultural Britain and global communities" (www.gazellemedia.co.uk) and aims to promote cultural diversity. The focus on Islam is mainly determined by the production team's interests and backgrounds. Most of its team are Muslim and interested in the topic of Islam in the west. Professionally trained, they have worked for major media companies, including the BBC and Channel 4 television, and have researched these issues before. For instance, executive director Navid Akhtar was a specialist advisor to BBC Radio 5 Live after 9/11. The team's expertise in arts, music, culture, religion and history goes beyond a purely subcultural focus. Gazelle Media also places an emphasis on Muslim youth culture in several ways: programmes for television and online broadcasting have included features on musicians, Sami Yusuf and Mohammed Yahya, as well as a film on British Muslims, celebrating the younger generations for comfortably identifying with both being British and Muslim. The company also offers media awareness training on Islam and creative expression, media literacy and production skills, and Islamophobia. For them, the latter is mainly the result of biased media coverage, particularly frustrating young Muslims, which may in some cases even contribute to radicalization. The courses are tailored to young people, schools, teachers, community organizations, journalists and members of ethnic minority groups. Finally, developing and promoting Islamic youth culture also counts among one of Gazelle Media's aims.

There are several other approaches to refashioning Islam for a new media age. A British medical student writes a popular blog in which she presents and comments on Islamic clothing (www.hijabstyle.co.uk). Although not a designer, she creates and proposes styles by combining non-Islamic fashion items with the hijab and in layers to achieve a full covering with style. The magazine *The Revival: Voice of the Muslim Youth* (first in print and online formats, from 2010 online only: www.therevival.co.uk) caters especially for young British Muslims. Articles address spiritual topics in ways that relate to young people, and demand, for example, that British imams acquaint themselves more with the particular issues facing young people today. Cover stories have included teenage pregnancy and drug abuse, to provide advice. The magazine also runs a popular column with two cartoon characters, Ali and Jamal: Jamal is characterized as the "good guy" who prays five times a day, knows the Quran by heart, never goes clubbing and who finally leaves for Egypt to pursue religious studies. Ali is portrayed as the "bad guy" with many girlfriends and little knowledge of religion; he drinks alcohol and eventually ends up in prison for drug dealing. It is overtly educational and, as with many examples of Islamic youth culture, the perceived "coolness" does not contradict a generally conservative outlook.

Similarly, according to their website, Muslim Youth Net provides the "coolest online space for Muslim youth" (www.muslimyouth.net). Like Waymo, it features the option to create a personal profile to participate in forum debates, upload videos and photos, or participate in polls. The organizers also run several campaigns to engage young Muslims; for example, the campaign "Faith Through Your Lens" was a photography competition on how young Muslims perceived religious beliefs, and the "Reelhood" campaign collected their views on the key issues they faced as young Muslims in Britain, which were then published on the website (www.muslimyouth.net/campaigns). The emphasis on offering support is underlined by the fact that the website organizers also run the Muslim Youth Helpline, an anonymous counselling telephone service specializing in the problems of young Muslims who might not feel sufficiently understood by other providers.

Finally, the website Muxlim (www.muxlim.com) promotes "Muslim Lifestyle" by collecting and reporting on Islamic products from around the globe, ranging from clothing lines to comics, music, news and businesses. Based in the UK and Finland, it was founded by a young Finnish Egyptian businessman who is active globally. As with the successful British magazine *Emel* (www.emel.com), his focus is on Islamic lifestyle regardless of age group, but the website's design and many of its featured topics and companies appeal especially to young people, in particular to the more affluent with an international perspective.

As the above examples have outlined, the Internet is an important tool for the dissemination of products and messages. Other notable examples include fashion designers Elenany (www.elenany.co.uk) and Styleislam (www.styleislam.com), who market and sell their clothes online, or rap artists Ammar114 (www.ammar114.deq) and Poetic Pilgrimage (www.myspace.com/poeticpilgrimage) who either have their own website or are active on MySpace, Facebook and Twitter. In many of these cases, new media platforms are useful tools to support product distribution, but less to facilitate an exchange with users of the Web 2.0 platforms. Evidence suggests that such exchanges still take place even more face to face, rather than virtually, through youth meetings or in everyday life. Still, this use of new information technologies highlights the growth of new virtual spaces as outlets for an Islamic youth culture.

Islam within Web 2.0 technology

The various new information technologies with a specific focus on young Muslims in the west cater to a variety of demands and desires, and, as Mandaville (2001, 183) has highlighted, demonstrate the plurality of Islam. Some of these websites report on and promote a certain genre, such as comedy at A part ça tout va bien. Local or global community news, including Muslim lifestyles, is shared through Waymo or Muxlim. In the first instance, these technologies contribute to the promotion of a sense of community. At the same time, some seek to educate, as in the control of content on Waymo, or the demonstration of Islamic humour by A part ça tout va bien. These differences suggest various levels of consumption and participation. Indeed, some outputs, such as video sketches or a fashion blog, are there to be consumed, but do not invite online interaction. They can anticipate a new style or identity shift, and their creators can hope for them to be adopted, or they can promote currents they observe and support. Others, like Waymo and MuslimYouthNet, supply a framework populated by user-generated content. To a certain extent, this also prescribes a desired identity by employing particular designs, mottos and images, and by controlling uploaded content. But there is great scope for participation in new media and Web 2.0 facilities, with ideas flowing back and forth between the website creator and the user. In this respect, new technology products are continually changing rather than "final", and created by many participants.

At the same time, the results of "offline" involvement through, for example, media training – provided, for instance, by Gazelle Media and MuslimYouthNet – are displayed online.

The heightened activity of young Muslims, encouraged through new information technologies, as well as by the nature of the products, are inherently hybrid. In global postmodern cultural identity politics, hybridity is associated with a mixing of different cultures; thus the process of hybridization contributes to the "making of global culture as a global mélange" (Pieterse 1993, 60). However, not only cultures from places that are far apart are blended; on a local level, hybridity denotes inversions, juxtapositions and fusions of "objects, languages and signifying practices from different and normally separated domains" (Werbner 1997, 2). Thus, it includes the possibility of mixing dissimilar elements that do not usually overlap. New information technologies with an Islamic message consciously bring together elements from diverging domains. In the cases discussed, the hybrid fusion is deliberate and intentional, as opposed to fusions that have taken place without reflection and by unconscious cultural adaptations throughout history (Werbner 1997, 5). It brings together elements that are widely regarded as incompatible. There is of course an obvious political dimension, seeking to challenge and overthrow the public discourse of what is considered the norm and against which difference is measured. Stuart Hall considers cultural identity a matter of proclaimed "unities" in a "play of power and exclusion" (Hall 1996a, 5), a state that can be opposed if identities are understood as a process of "becoming rather than being" (1996a, 4). The use of various historic, linguistic and cultural resources leads to a hybrid creation of identities, rather than the continuation of an essentialized identity. Cultural identity, according to Hall, is always hybrid rather than fixed (Hall and Chen 1996, 504). In the case of black popular culture, he perceives "no pure forms at all", as these forms are invariably the result of adaptions, confluences and negotiations across cultural boundaries (Hall 1996b, 474). However, it is not just important to think of cultural identity as a process of becoming, but rather as active production. Hall's understanding of diaspora as a constantly new production and reproduction of identities is relevant in the context of Islam in Europe: "Diaspora identities are those which are constantly producing and reproducing themselves anew, through transformation and difference" (1990, 235). Hall rejects a definition of diaspora derived from purity and essence, and instead underlines that it is defined by heterogeneity, by identities based on difference. One might assume that in the diaspora situation cultural identities are thought to be produced regardless of difference, or in an attempt to hide this difference, but what Hall emphasizes is that the diasporic "conception of "identity" […] lives with and through, not despite, difference" (1990, 235). It is this productive engagement with cultural difference that he calls hybridity, a process through which innovative and mixed – or hybrid – forms of (popular) culture are created. In the examples discussed above, new information technologies are expressions of such hybrid identities, as they reveal the appropriation of British and Islamic, subcultural and religious characteristics for the creation of a Muslim youth-specific identity. They underline the reality of a hybrid identity, which dominant society has so far mainly considered in separate and seemingly incongruent parts.

The ummah, the global community of Muslims, is often invoked as being further united by the use of global media and the Internet. Mandaville (2001, 183) also speaks of the Internet as an opportunity for those who feel excluded from dominant society to find like-minded individuals, which leads to a new form of community experience or to "reimagined Islam" (183). I would argue, however, that involvement with new information technologies remains rather local and, despite its global connectivity, actually strengthens a community feeling on a local level. For example, Waymo supports the development of a German Islam, and indeed references to anything German – language, local events, customs, politics – are significantly more present than references to Muslim life in Islamic countries. Thus, the online space is not so far removed from the actual space in which the user interacts, but acts more as a filter of the environment to find like-minded people locally.

In recent years, some researchers have interpreted the presence and influence of new information technologies as causing a substantial change in Islam itself. Mandaville (2001, 184) claims that they enable new interpretations that contribute to a "remaking of Islam". Saminaz Zaman (2008, 469–471) sees new religious authorities on the horizon due to the Internet, because many users search online for religious advice. This can take the explicit form of online fatwas, which Jocelyne Cesari (2004, 111) also points to in her concept of "virtual Islam" , or appear as religious advice in the more playful education of Waymo or *The Revival*'s website. Zaman (2008, 470) also points out that the Internet "stimulates nostalgia as much as newness" , if used to listen to the call for prayer or Quranic recitations, and warns against overrating the innovative character of new information technologies. The examples presented here, however, show that Islamic youth culture expresses itself through new information technologies, but that the medium is in most cases more than a mode of conveying particular content. Thus, youth-orientated media platforms create a space for young Muslims to interact with each other and with Islam – for instance, by learning about it or forming an opinion and possibly an identity based on elements of Islam.

The examples of new information technologies in Islamic youth culture suggest that there is a strong concern with the everyday environment of Muslims who produce and consume them. The producers' aim may therefore not just be the development of a purely virtual culture. While the products themselves do not reveal sufficiently the purpose of their development, their creators have various clear aims beyond the simple fashioning of new information technologies, and they have given manifold accounts of their activities in in-depth interviews.

Virtual support for local commitment

During the interviews I conducted for this research (see Appendix 2), the content producers of Islamic media and new information technology platforms revealed a range of motivations for their engagement in their activities. Within the diversity of accounts, two main groups emerged, both of whom were very active in their respective national contexts. The first group dealt with society as a whole, and the second group was mainly concerned with young Muslims. Of those interested in wider society, French comedian Hilal,[3] for instance, expressed a strong urge to foster understanding between Muslims and non-Muslims in France, using comedy to defuse an often tense and volatile relationship. Despite his strong religiosity, he considered himself more a fellow human and a French citizen than a Muslim and emphasized that he condemned any sort of sectarianism – an accusation he often felt confronted with as a French Muslim. A British respondent, Nour, favoured a more subtle and complex engagement with wider society through his media productions, which dealt with the traits that Islam and western culture shared through, for example, abstract art. He argued for a more sophisticated way of fusing Islam and the west than that in most examples of Islamic youth culture, which he found unimaginative and superficial. Instead, he underlined his firm belief in shared, deep-rooted cultural ideas. In his opinion, both Muslims and non-Muslims could profit from the knowledge of such common roots in their everyday interaction in British society.

In contrast to the majority of producers with a localized focus, businessman Mansour was internationally active. As the founder of a Muslim media platform, he aimed to counter the misrepresentation of Islam in the mainstream media in Britain and to provide information about Muslim lifestyle and consumption. Mansour negotiated with British government officials to improve the representation of Muslims in the media on a global scale. Although he was less concerned with particular national or local settings, much of his experience drew on encounters between Muslims and non-Muslims, especially outside the Islamic world.

Those among the producers who were mostly concerned with young Muslims mainly addressed topics of empowerment and identity. In Germany, the most prominent activist was Hamid, a founder of an online platform, whose motivation was to provide a tool for young Muslims to help them develop and exchange a culture of their own. By attracting the more religious element of young Muslims and giving them an online environment, he sought to empower them to actively shape their identity, unencumbered by elders or non-Muslims. Although Hamid's approach could certainly be regarded as "top-down", since he sets the framework and controls discussions, his aim, rather, was to assist young people by providing them with the space and freedom to develop their own thoughts, in a forum where they could express their concerns.

In Britain, Mahmoud, co-editor of a Muslim youth magazine, wanted to help develop a new youth identity through a fusion of British and Islamic traits of which young Muslims could be proud. Witnessing the difficulties of disadvantaged young Muslims in the north of England, who felt isolated from and rejected by British mainstream culture, he wanted to highlight through his magazine pathways for a positive identification with Britain.

Rasha, also living in the UK, who worked for a helpline and webspace for young Muslims, similarly addressed the idea of empowerment. She placed young people at the centre of her narrative and emphasized their ownership of the website's user-generated content. In her work, she did not want to provide solutions, but rather empower young people to reach their own decisions. Jamila, who wrote a fashion blog, was one of the few to reach out to a broader international audience. She longed for normality, and liked being able to "feel like everyone else". Instead of stressing the difference between Islamic dress and other styles, she emphasized her interest in fashion that she shared with other Muslim women and had in common with many other British women. Her blog was directed not at non-Muslim British society, but at conservative Muslims. Her motivation in writing it was that she felt at home in British society. She emphasized her understanding of normality – being able to share everyday experiences of mainstream society, such as a liking for fashion – and demanded that it be accepted by conservative Muslims who were still sceptical of the combination of aesthetic fashion and Islamic covering.

Such empirical accounts and examples illustrate how new information technologies in some ways challenge received notions of Islam and youth culture. These subcultures display a high degree of religiosity and translate religious traditions into forms and behaviours associated with contemporary ideas of "coolness". This contrasts with other subcultures and new information technology products that thrive beyond religion. The trend also deliberately challenges unfavourable and stereotypical mainstream media representations of Islam in the traditional British press (Yaqin 2010, 229–230, 238;

Morey 2010, 252–254). The adoption of new information technologies, then, stands for a form of Islam that is lived and practised in youth-specific, non-traditional ways and in itself challenges other, more orthodox forms of Islam.

A transnational consciousness?

The use of new information technologies in Islamic youth culture suggests a transnationally active movement. Indeed, the issues negotiated on Internet platforms or in comedy sketches – such as identity quests or Muslim/non-Muslim relations – recur in British, French and German settings alike. A constant reference to the global ummah is prevalent among the followers of Islamic youth cultures. The desire to exchange information about Islam by using new forms of online communication, particularly the utilization of the Internet, has the potential to break down previous barriers.

Surprisingly, however, there is currently little evidence of a transnational or even a global orientation. New information technology producers rarely make references to similar formats in other countries and, when asked, were not usually aware of them. One of the main components seems to be the focus on the local conditions of young Muslims and the urge to improve them. Language plays another role: instead of Arabic or any other language from an Islamic country, the young European Muslims use German, French and English. This does does not mean that communication between countries would be impossible, as media users or producers could use a common language such as English, but the evidence suggests that exchanges between these countries are not widely practised.

These observations highlight that there seems to be a stronger necessity to use new information technologies to support local issues, rather than to make use of their transnational potential. The global dimension is certainly part of Islamic new information technologies – but ultimately the content and format are local, determined by everyday life in respective national contexts. Questions, therefore, remain about the relationship between the global and local. Undeniably, new information technologies in Islamic youth culture have a significant transnational dimension in their appropriation of global youth cultures, which links them to some extent to a wider shared cultural history, norms or ideas (Therborn 2011, 117, 160). The fact that a global religion in a specific geopolitical setting is involved here adds to the transnational character. Finally, the Internet provides global connectivity. As a result, new Islamic information technology products will be trans-cultural or multicultural in many ways.

At the same time, local variations and specificities are equally to be expected. Although it is acknowledged that globalization does not simply result in homogenization (Lukose 2005, 931), youth cultures are also seen as strongly dependent on national and social frameworks. Local meanings are added to globally travelling cultures, especially to commercialized North American ones. In turn, global trends and problems are interpreted nationally (Skelton and Valentine 1998; Androutsopoulos 2003; Wise 2008). The appropriation of global flows of culture for local purposes depends on local conditions and on constraints experienced in different countries. As revealed by the producers' accounts, globalized subcultural traits are used, but are developed for immediate local needs.

New information technologies are used for virtual communication, but, despite the opportunities this offers, there does not appear to be a desire for exchange with similar actors in other European countries, nor an urge to act as part of a unified youth movement. Instead, the technology is used to facilitate exchange between those who share a similar localized situation, where new technologies offer a virtual environment of support that does, however, not act as a substitute for non-virtual interaction. While on the surface Islamic youth culture's presence in new information technology may appear disjointed, it can be regarded as a collective response to post-9/11 and 7/7 environments. It addresses the pertinent issue of regaining control over labels, categorizations and stereotypes ascribed to Muslims by dominant society. At the same time, differences between countries, as displayed by the respondents (for instance, a more positive perception of society by British interviewees as compared to participants from France and Germany), point to the great influence of national contexts that are determined by the integration models and differing notions of citizenship, media and public opinion in the respective countries.

Conclusion

The use of new information technologies in Islamic youth cultures takes on a variety of forms. Users and producers attach their own specific meanings to them, ranging from entertainment to religious education and solutions to social problems. While many of these negotiations take place online, in a virtual space that brings people of similar experiences and backgrounds together, the issues they debate are rooted in everyday life. The platforms developed by new information technologies, then, help enable participants to communicate, and to take an innovative view of religion and society. However, they do not act as a substitute for reality, nor do they generate a transnational movement. Producers and consumers of new Islamic information technologies are more concerned with the situation in their own country and with ameliorating that situation – especially for young Muslims – than with utilizing the capacity of the Internet to transcend borders. Despite the ways in which they appropriate global flows of technology and culture, the ideas behind most of the Islamic platforms discussed in this article remain specifically local, and deal primarily with the situation of Muslims and non-Muslims in their respective European countries.

Notes

1. The methodological approach of the full study was based on an ethnographic and reconstructive approach, using a variety of collected data as a basis: youth cultural artefacts such as religious rap, street-wear with Islamic slogans, comedy sketches and Internet facilities for the exchange of young Muslims' issues; participant observations among consumers of Islamic youth culture; and 32 qualitative in-depth interviews with French, German and British producers of Islamic youth culture from different subcultural genres.
2. A webisode, a portmanteau word of web and episode, denotes a short video released on the Internet instead of television and is usually part of a series.
3. All names have been changed. A list of the interviews can be found in Appendix 2.

References

Androutsopoulos, Jannis, ed. 2003. *HipHop: Globale Kultur – Lokale Praktiken*. Bielefeld: Transcript.

Boubekeur, Amel. 2005. "Cool and Competitive: Muslim Culture in the West." *ISIM Review* 16: 12–13.

Cesari, Jocelyne. 2004. *When Islam and Democracy Meet: Muslims in Europe and in the United States*. New York and Basingstoke: Palgrave Macmillan.

Gerlach, Julia. 2006. *Zwischen Pop und Dschihad. Muslimische Jugendliche in Deutschland*. Berlin: Links.

Hall, Stuart. 1990. "Cultural Identity and Diaspora." In *Identity: Community, Culture, Difference*, edited by Jonathan Rutherford, 222–237. London: Lawrence & Wishart.

Hall, Stuart. 1996a. "Who Needs 'Identity'? In *Questions of Cultural Identity*, edited by Stuart Hall and Paul du Gay, 1–17. London: Sage.

Hall, Stuart. 1996b. "What is This 'Black' in Black Popular Culture? In *Stuart Hall: Critical Dialogues in Cultural Studies*, edited by David Morley and Kuan-Hsing Chen, 468–478. London: Routledge.

Hall, Stuart, and Kuan-Hsing Chen. 1996. "The Formation of a Diasporic Intellectual: An Interview with Stuart Hall by Kuan-Hsing Chen." In *Stuart Hall: Critical Dialogues in Cultural Studies*, edited by David Morley and Kuan-Hsing Chen, 486–505. London: Routledge.

Herding, Maruta. 2013. "Inventing the Muslim Cool: Islamic Youth Culture in Western Europe." Bielefeld: transcript.

Lukose, Ritty. 2005. "Consuming Globalization: Youth and Gender in Kerala, India." *Journal of Social History* 38 (4): 915–935.

Mandaville, Peter. 2001. "Reimagining Islam in Diaspora: The Politics of Mediated Community." *International Communication Gazette* 63 (2-3): 169–186.

Mandaville, Peter. 2009. "Hip-Hop, Nasheeds, and 'Cool' Sheikhs: Popular Culture and Muslim Youth in the United Kingdom." In *In-between Spaces: Christian and Muslim Minorities in Transition in Europe and in the Middle East*, edited by Christiane Timmerman, Johan Leman, Hannelore Roos, and Barbara Segaert, 149–168. Brussels: Peter Lang.

Morey, Peter. 2010. "Terrorvision." *Interventions: International Journal of Postcolonial Studies* 12 (2): 251–264.

Nederveen Pieterse, Jan. 1993. "Globalization as Hybridization." In *Global Modernities*, edited by Mike Featherstone, Scott Lash, and Roland Robertson, 45–68. London: Sage.

Skelton, Tracey, and Gill Valentine, eds. 1998. *Cool Places: Geographies of Youth Cultures*. London: Routledge.

Therborn, Göran. 2011. *The World: A Beginner's Guide*. Cambridge: Polity.

Werbner, Pnina. 1997. "Introduction: The Dialectics of Cultural Hybridity." In *Debating Cultural Hybridity: Multi-Cultural Identities and the Politics of Anti-Racism*, edited by Pnina Werbner and Tariq Modood, 1–26. London: Zed.

Wise, Macgregor J. 2008. *Cultural Globalization: A User's Guide*. Carlton: Blackwell.

Yaqin, Amina. 2010. "Inside the Harem, outside the Nation." *Interventions: International Journal of Postcolonial Studies* 12 (2): 226–238.

Zaman, Saminaz. 2008. "From Imam to Cyber-Mufti: Consuming Identity in Muslim America." *The Muslim World* 98 (4): 465–474.

Appendix 1: Summary list of websites referred to

(All online resources were last accessed on 4 March 2012 unless stated otherwise.)

A part ça tout va bien: http://www.apartcatoutvabien.com.

Ammar114: http://www.ammar114.de.

Elenany: http://elenany.co.uk.

Emel Magazine: http://www.emel.com.

Hijab Style Blog: http://www.hijabstyle.co.uk.

Islam School Welkoum: http://www.islamschoolwelkoum.com.

Muslim Café: http://www.muslimcafe.tv 14/09/2008,

Gazelle Media: http://www.gazellemedia.co.uk/projects/view/muslimcafe-tv

Muslim Youth Helpline: http://www.myh.org.uk.

MuslimYouthNet campaigns: http://www.muslimyouth.net/campaigns.

Muxlim: http://muxlim.com.

Poetic Pilgrimage: http://www.myspace.com/poeticpilgrimage.

SaphirNews: http://www.saphirnews.com.

Styleislam: http://styleislam.com, http://blog.styleislam.com. (See also Styleislam on Facebook and Twitter.)

The Revival: http://www.therevival.co.uk.

Waymo: http://waymo.de.

Appendix 2: Interview material

Interview #	Fictional name	Sex	Country	Date of interview
A5	Hamid	M	Germany	14 Nov. 2008
B1	Hilal	M	France	7 Feb. 2009
C10	Mansour	M	UK	10 Aug. 2009
C11	Jamila	F	UK	13 Aug. 2009
C12	Mahmoud	M	UK	01 Sept. 2009
C13	Rasha	F	UK	9 Sept. 2009
C15	Nour	M	UK	19 Jan. 2010

New media beyond neo-imperialism: *Betty Boop* and *Sita Sings the Blues*

Sandra Annett

Wilfrid Laurier University, Waterloo, Ontario, Canada

This essay argues that theories which frame media globalization as either Disney-style neo-imperial domination or as radical technopolitics fail to account for the complex network of exchange that takes place in animated media. Instead of framing media use in terms of "top-down" versus "bottom-up" activity, this essay demonstrates how animators and audiences in different national and cultural contexts may inhabit multiple positions, entering into fraught, yet often productive, relations of complicity and collaboration through different media technologies. To that end, it highlights two particular historical moments at which emerging media became linked to cross-cultural networks of exchange. The first moment is that of Betty Boop's birth in 1930, which coincided with Hollywood's rise to prominence in the global distribution of sound film as far afield as Japan. Betty Boop's reception in Japan is considered through an examination of 1930s Japanese advertising documents and the parodic short films of animator Ōfuji Noburō. The second moment is the digital shift of the early 21st century, illustrated in Nina Paley's 2008 online *Ramayana*-based musical *Sita Sings The Blues*, and its reception among diasporic and national South Asian audiences. Though produced decades apart, these films are linked by concerns around nationality, ethnicity, gender, and cross-cultural exchange, demonstrated in their repeated use of a single core image: the exoticized and eroticized female singer protagonist. This problematic (yet potentially powerful) figure is analysed in both cases as a focal point for neo-imperial complicity and collaborative reinterpretation among creators and viewers of global media.

For scholars such as Armand Mattelart there is nothing *new* about today's media or imperialism. "The conquest of the cyber-frontier", Mattelart argues, "is a sequel to the grand technological narrative of the conquest of space" (2003, 1). In his view, just as in the 19th century, when London was the undisputed hub of the transcontinental network of underwater cables, today the United States has become the nodal point through which Net users from less-developed countries must go in order to connect with each other (148). Territorial colonialism has given way to virtual spaces of power, but the hierarchies of centre and periphery created by unequal access to information technologies remain. This kind of mediated neo-imperialism is often thought to go hand in hand with cultural imperialism, in which American media giants control not just the access ports, but also the content of what the world watches. In the 1970s Walt Disney's comics and animation were famously criticized by Mattelart and his co-writer Ariel Dorfman as examples of cultural imperialism in Latin America (Dorfman and Mattelart 1975).

Theorists of media globalization such as Lee Artz have likewise condemned the "corporate media hegemony" (2003, 26) of "Disney's menu for global hierarchy" (2005, 75) well into the 21st century. "Disneyfication" stands as a catchword for American media neo-imperialism: a form of "networking the globe" that ensnares media consumers around the world as much as it interconnects them.

In response to these "(gloomy Marxist notions of) Westernization or American imperialism" (2008, 11), as Revathi Krishnaswamy calls them, many scholars of postcolonialism and globalization have turned to ideas of "subversive consumption" (11) or "active audiences" to formulate resistance to neo-imperial domination. John Tomlinson's *Cultural Imperialism* takes Dorfman and Mattelart to task for ignoring the crucial question of "*the relationship between text and audience*" (1991, 44; emphasis in original). He cites Ien Ang's studies of the diverse audiences of *Dallas* as evidence that "audiences are more active and critical [...] and their cultural values more resistant to manipulation and 'invasion' than many critical media theorists have assumed" (50). More recent studies of online activist communities even turn new media against neo-imperialism, to the point of claiming, as Richard Khan and Douglas Kellner have, that a "radical technopolitics [...] can liberate humanity and nature from the tyrannical and oppressive forces that currently constitute much of our global and local reality" (2005, 94–95).

Centre–periphery models of neo-imperialism can fail to acknowledge the multiple positions inhabited by animators in their own countries, and overlook the complex networks of exchange that exist between media creators and spectators in different nations. As I shall demonstrate, even works of "American" animation may in fact be transnational or diasporic in their production. Likewise, the intended audiences for these works in countries such as Japan and India are far from homogeneous national masses, but include individuals who respond creatively to globally distributed animation from their own historical, social and technological experiences.

And yet, just because animators and audiences are active participants in transnational networks, it does not make them media revolutionaries standing in clean opposition to all dominant discourses. My argument is that animators, their texts and their audiences do not stay neatly within the bounds of a single national or cultural ideology, but neither can they fully escape their positionings within the economies of multinational capitalism and the politics of identity formation across countries, ethnicities and genders. Instead of framing animation in clearly opposed terms of neo-imperial domination versus new media resistance, I shall examine how animators and audiences in different national and cultural contexts enter into fraught, yet often productive, relations of complicity and collaboration through different media technologies.

Many media platforms today show evidence of the two intertwined kinds of relations described here: relations of *complicity* which subtly reaffirm the power imbalances of (neo-)imperial discourses, and relations of *collaboration* which generate conversations across difference, exchanges that are mutual yet asymmetrical. My essay will address such relations, going beyond the discourses of "Disneyfication" to consider an animated character just as renowned as Disney's Donald Duck, but which presents different challenges at the levels of postcolonial and feminist media criticism: Betty Boop. Since her debut, the Fleischer Brothers' Betty character has served as a globally consumed image of the exoticized and eroticized female singer. Her image functions as a perfect commodity in world markets, but also as a site for contention and parody, as is seen in adaptations from the 1930s to the present day. Rather than providing a teleological, determinist narrative of Betty's "evolution", however, this essay highlights particular *moments* of historical transition: singular points at which the image of the female singer became linked to key changes in media technologies and modes of cross-cultural interaction.

Betty Boop's birth in 1930 marks the first moment, coinciding with Hollywood's rise to prominence in the global distribution of sound film as far afield as Japan. The second moment is the digital shift of the early 21st century, as seen in Nina Paley's 2008 web-based musical *Sita Sings the Blues* and its reception among diasporic and national South Asian audiences. In both instances, the exotic and erotic singer appears as a figure poised between imperial complicity and collaborative reinterpretation. Her self-reflexive "performances" are linked by intersecting concerns around nationality, ethnicity, gender, and cross-cultural exchange. But the approaches animators take to these common tropes are reshaped by their historical contexts and technological platforms: 20th-century film and the 21st-century Internet. In this way, the case of Betty Boop's travels reflects deep changes in how animated media is created and how audiences interact with it and each other across cultural and national boundaries. By comparing how Betty Boop has crossed between platforms and nations through the historical shift of silent to sound film and the current shift from analogue to digital visual media, we may learn to recognize the risks of neo-imperial complicity and develop the potentials of transnational collaboration that arise in times of media transition.

Betty Boop in America and Japan: a case of "imperialist internationalism"

In order to explore how animation can support or complicate neo-imperialism, it is first necessary to look at the history that has shaped our basic attitudes towards animation. Beginning in the late 1920s and arguably continuing until the present day, the short and feature-length animated films of Walt Disney have acted as the canonical standard against which animated works are judged, either positively or negatively. But while Disney may be the most recognizable of the Golden Age animators today, the Walt Disney Studios were far from unrivalled in their time. Equally as well known and influential in the 1930s–1940s were the Fleischer Studios, founded by Max and Dave Fleischer.

The Fleischer brothers were the children of a Polish-Jewish immigrant family who moved to New York in 1887, when Max was 5 years old. His younger brother Dave was born in 1894 in New York. The pair began their professional careers there in 1918 when they produced their first silent cartoon, the *Out of the Inkwell* series featuring Koko the Clown. They introduced Betty Boop as a side character in 1930; by 1932, she was a famous cartoon starlet in her own right. Early on, the *Betty Boop* series appealed to adults as much as to children, with its sexy heroine and "gags built on urban and industrial experience, a fantasy world of neighbourhoods, sweatshops, pool halls, Coney Island rides, and [...] Manhattan vaudeville" (Klein 1993, 62). As well as attracting a general urban audience, the shorts were grounded in the cultural climate of New York's Lower East Side Jewish immigrant neighbourhood, reflected in their use of Yiddish-language humour and film conventions. As Amelia S. Holberg argues in her article "Betty Boop: Yiddish Film Star", along with the language of the Hollywood-style musical cartoon,

> Betty's cartoons also spoke the language of the Yiddish cinema. That language included not only bits of actual Yiddish but also references to the themes of the Yiddish cinema and the lives of working-class Jews jammed together in tenements on the Lower East Side. [...] the Fleischer cartoons are a prime example of a unique moment in American cinema

in which a product aimed at a mass audience also reflected the concerns and culture of another cinema audience altogether – the audience for the alternative Yiddish cinema. (1999, 302)

Already, the Fleischers' works diverge from the homogeneous, homogenizing American national culture assumed in critiques of "Americanization", representing instead the diasporic experiences of immigrants.

That is not to say, however, that the Fleischers were exemplars of multicultural, multi-ethnic empowerment. Holberg notes that the success of Jewish film-makers often depended on the exclusion or caricature of fellow immigrants, such as Chinese workers, and of black people, who were almost always cast in the roles of American jazz men or African cannibals in Fleischer cartoons. Adding gendered stereotypes to the mix, Betty herself was sometimes painted as a sensual "ethnic" character. In the short *Betty's Bamboo Isle*, her skin was darkened and she was dressed in a skimpy grass skirt in order to perform a dance traced from the filmed movements of a Samoan dancer named Miri. Joanna Bouldin argues that Betty's representation here draws on the trope of the "ethnographic body", exemplified in the spectacle of the exotic woman caught on film for the educational pleasure of an assumed white male audience (2001, 52–53). In the early 1930s, then, exotic and erotic imagery played out on the animated screen in ways that disturbed but also reinforced the hegemonic imperial discourses underlying mainstream Hollywood cinema.

Still, as the Depression deepened and the social climate grew harsher, films like *Betty's Bamboo Isle*, which drew on the "exotic erotic" formula, soon became subject to a growing moral panic surrounding Hollywood film. This panic culminated in the creation of the "Hays Code", a motion picture production code designed to censor anything, including nudity, suggestive dancing and interracial relations, that might "stimulate the lower and baser element" in audiences (Hays Code 1930, n.p.). By 1934 the Code was regularly enforced, so that within four years of her debut Betty's flapper days came to an end, leading to a drastic redesign (Hendershot 1995, 120). The Fleischers were thus forced to seek new ways to enhance their star's appeal.

One of these ways was to turn to the international market. According to animator Myron Waldman, the Fleischers became aware that Betty Boop was popular in Japan, and decided to create a short "designed to appeal to the Japanese market" (Dobbs 2006, n.p.). This was *A Language All My Own*, which features Betty performing the title song about how her catchy tune brings people around the world together. After singing for a cheering New York audience, Betty sets off for the Land of the Rising Sun, depicted literally with an emblematic sunrise over Mt Fuji. The opening seems like a perfect set-up for the kinds of racial caricature comedy seen in *Betty's Bamboo Isle*. In this case, however, the Fleischers were deeply concerned not to offend their Japanese fans. As a result, the members of Betty's Japanese audience are not depicted as the usual pigtailed pan-Asian grotesques, but as more proportionate adult figures with detailed kimono designs and hairstyles – though still rather bucktoothed and hardly individualized. Even more surprising, Betty sings not only in English, but also in Japanese. Waldman recounts that the staff consulted with Japanese exchange students in America on the lyrics and on Betty's dance, to be sure her body language and gestures would not offend anyone in Japan. In this case, animating became, just for a moment, a collaborative process that valued the input of those it sought to represent, resulting in a fascinatingly hybrid work.

And yet even this sort of "hybridism" echoes Betty's earlier performance of the exotic and erotic "ethnographic body". Upon her arrival in Japan, Betty appears on stage in a flourish of unfolding fans and begins a performance in which she physically enacts the "east" and the "west", reflecting both in telling ways through her body and her lyrics. When she sings the line "If you're near or far / doesn't matter where you are", she sways, loose and sinuous, to the tune of "The Streets of Cairo", a piece made famous by its use in the sensationalized belly dance performances of "Little Egypt" at the 1893 Chicago World's Fair (Carlton 2002, 69). But when Betty sings her next line, "Song's in ev'ry land o'er the ocean", she stands at attention and salutes to an American-style march. Though much more subtle than the exotic eroticism of *Bamboo Isle*, the overall combination of music, images and words in *A Language All My Own* still suggests that to be "far" is to be embodied as a sensual oriental woman, while the universality of song is uprightly western.

What is more, it is the catchphrase that made her famous in America, her "boop-boop-a-doop," that is "known in every foreign home". Betty has her Japanese fans repeat this line and they chime in happily with the refrain. The depiction of Japanese audiences as ready imitators paints a screen-dream of Japan as a land of compliant consumers ready to sing along to western tunes. Betty's performance and the audiences' "participation" in this short subtly reveal the Fleischers' complicity in Orientalist conceptions of bounded, embodied national identity, on which the cartoon's attempt to build international relations was founded. In this way, the short can be seen to embody an "imperial internationalism" that promises a language of easy connection, yet is underpinned by persistent colonial and imperial discourses.

However, as Tomlinson (1991) has noted, when considering the issue of "cultural imperialism" in media it is just as important to look at the actual conditions of distribution and reception as at production. If the content of Betty Boop cartoons can be seen as hegemonically Orientalist, a closer look at how the image of Betty was framed and reworked in Japan reveals a situation that, while still problematic, is much more complex than the unilateral dominance of Japanese culture by international–imperialist ideologues.

First, there is the issue of distribution and promotion. In the early 20th century, the exclusive distributor for Fleischer Studios was Paramount Pictures, which also handled Disney films in Japan. Paramount began marketing American films from their Tokyo office as of 1930 (Anderson and Richie 1982, 75–76). They focused particularly on the new technology of sound film, an area where American imports initially held a 90 percent market share in Japan (Thompson 1985, 143). Paramount's Tokyo branch also actively promoted Betty Boop talkies, placing full-page ads with lists of the latest Fleischer imports in the major Japanese film magazine *Kinema Junpo*. Among these is an advertisement from the 1 November 1935 edition for *A Language All My Own*, retitled *Japan Visit* (*Nihon hōmon*). Taking up half the page, it features a stylish line-art image of Betty Boop flying her plane over Mt Fuji (Figure 1). The accompanying text proclaims Betty to be the "Queen of Popularity", and provides the following puff: "Paramount Cartoon Studios' masterpiece! Betty Boop, a cartoon goodwill ambassador between Japan and America, visits Japan and sings in Japanese in this splendid masterpiece! Betty's 'Japan Visit' " (Tsutsui 1992, 225).

Here, the attempt at international communication seen in the Fleischers' short is brought out even more strongly than in Waldman's own statements. Rather than speaking of markets, it uses the language of diplomacy and international relations, evident in the phrase "cartoon goodwill ambassador" or "*nichibei shinzen no manga shisetsu*"

Figure 1. Ad for Betty Boop's *Japan Visit* in *Kinema Junpo* magazine, 1 November 1935. Betty Boop TM Hearst Holdings, Inc./Fleischer Studios, Inc.

(日米親善の漫画使節), which suggests a government envoy or delegation (*shisetsu*) aimed at promoting Japan–US friendship (*nichibei shinzen*). The idea that film could be used as a political tool was far from foreign in Japan. Beginning in the late 1920s and early 1930s, there was a push among reform advocates for the formation of a national film policy, rooted in a "desire to promote Japanese films abroad as an intercultural exercise in mutual understanding" (Standish 2006, 140). In the advertising for *Japan Visit*, Betty Boop was subtly repositioned through issues of international relations that concerned (the more official parts of) the Japanese film world.

When Betty Boop entered Japan as a "goodwill ambassador", then, she was not entering a theatre full of quaint kimonoed figures eager to sing along as instructed, as in the cultural-imperialist dream, but a modern(izing) social field fraught with changing discourses regarding the role of cinema, the nation and international relations. Western works may have been popular, but they were not always passively consumed. In some cases, they also became part of the raw visual material used in animated film production. Indeed, by the time of Betty's fictionalized arrival in 1935, her image had already been taken up and transformed by those among the Japanese audience who were also film creators, such as Ōfuji Noburō.

Ōfuji himself was a film-maker poised between worlds. Born in Asakusa in 1900 and trained by pioneering animator Kōuchi Jun'ichi, Ōfuji was deeply inspired by both domestic film-makers and by the works of European and American animators (Yamaguchi and Watanabe 1977, 15–16). This background is evident in his animated short *Tengu Taiji*, or *Defeat of the Tengu*. There are at least two versions of this film in existence: a talkie version with no title cards, and a silent version with title cards meant to be interpreted by a *benshi* narrator; both are dated 1934. In both incarnations, Ōfuji visibly draws on the Fleischers' style to tell a fantastic period-drama story, displaying once again the intertwining influences of animated cinemas in international circulation.

Defeat of the Tengu opens with a little dog-boy named Heibei, who is on fire-watch duty when a black-feathered arrow shoots over his head, hits a wall and morphs into a grotesque face that laughs at him. The arrow was fired by a marauding gang of bird-like mythological creatures called *tengu*. The *tengu* miscreants break into a nearby geisha house to kidnap one of the women, squashing flat as a sheet of paper the man who tries to stop them. Heibei, who has been hiding all this time in his own apron, cries "Taihen da!" ("How terrible!"), and takes the flattened figure to the great Lord Hyōei. The squashed man, it turns out, was Hyōei's beloved uncle. Hyōei vows revenge, and, folding his uncle into an origami helmet, he rushes off after the *tengu* with theatrical gestures. Like the American animated shorts which spoofed feature melodramas, this

Figure 2.   Sir Hyōei in *Defeat of the Tengu*. *Japanese Anime Classic Collection* (public domain film).

film culminates in a parody of the popular *chambara* sword-fighting genre. In a wonderful mock-epic battle, enemies and heroes alike are sliced in half only to literally pull themselves back together. Hyōei may defeat the *tengu* army, but it is little Heibei who saves the samurai from the chief *tengu* by clipping off its famous long nose with a crab's claw. In narrative, the short is quite different from American animation, drawing on a domestic Japanese genre. But in visual style and gags, such as the surreally transforming arrow, it is highly reminiscent of Fleischers' cartoons. The true tip-off is the character design of Hyōei himself, who can only be described as Betty Boop in a top-knot. (Figure 2)

Defeat of the Tengu thus reads as a two-pronged reflexive parody. On the one hand, it makes fun of the live-action Japanese film genres of *chambara* and *jidaigeki* (the period piece). On the other hand, it plays on, or rather plays *against*, Betty's canonical appearances as an American film star by situating her in a markedly "Japanese" historical setting. In being recast as Hyōei, Betty no longer performs alluring Orientalist femininity, but a parodic martial masculinity asserted in overblown heroic gestures. It is worth noting that this kind of martial masculinity would soon become a feature of wartime propaganda films designed to shore up the Japanese Empire's colonial presence in East Asia, such as *Momotaro's Divine Ocean Warriors* (directed by Seo Mitsuyo in 1945). In this pre-war short, however, Hyōei's heroism could still be comically undercut, as he is saved from the chief *tengu* by little Heibei with a crab claw that, moments before, was pinching Heibei's bottom. In reflexively recasting Betty along lines of gender and genre completely different from those she was created to play and then *over*playing them, this short embodies the transformative power of local appropriations of "dominant" Hollywood film.

At the same time, such appropriations did not take place in an empty playing field free of all economic considerations. For instance, I have noted that *Defeat of the Tengu* was produced in both silent and sound versions. The film stood at the transitional point between earlier animation methods and technological changes that were taking place in film production as a result of foreign competition. The growing popularity of sound film was a major factor. According to Yamaguchi and Watanabe (1977), competition from American sound film caused great hardship for Japanese animators. Since American studios made most of their revenue from domestic American sales, they were able to mass-produce prints for overseas markets at a relatively low cost. Japanese animators, by contrast, worked under a craft system, in which a single artist such as Ōfuji formed his own studio and made cartoons with the help of a few apprentices, creating fewer prints. For them, sound recording was expensive; the production process took three to four times longer, and they could rely on only one source of revenue: the domestic Japanese market. On average, animators had to charge 1000 yen per one-reel short film to cover costs. Theatre owners increasingly refused to buy domestic shorts at that price, protesting that "for 1000 yen, we can get two Mickey Mouse talkies" (Yamaguchi and Watanabe 1977, 26). To be undersold in their own market was a dire blow for Japanese animators, as they were not able to export their animation along global trade routes that largely shipped finished films one way: from west to east.

The flood of Betty Boop products in Japan, to the detriment of domestic film, is a classic example of the kind of economic imperialism prevalent in many other areas of trade in the early 20th century. It cannot be called *complete* foreign domination or colonization, since profits from locally made Betty Boop products did not return to the Fleischers. Furthermore, American cultural hegemony was not total, but was continually being renegotiated within the complex fields of Japanese cultural identity and imperial ambitions. But the "Betty boom" did nonetheless result from a vast power imbalance in global trade which allowed American products to flourish in Japan while freezing Japanese creators out of the world market. In this way, the chances for *mutual* cultural exchange and collaboration were foreclosed in favour of a structure of imperial internationalism, in which many animators were unwittingly complicit.

Sita Sings the Blues: online collaboration and transnational frictions

As suggested so far, the key issue when considering media and (neo-)imperialism is not who creates or consumes media and who does not, but what structures exist to mediate the relations between differently positioned audiences and creators. This remains a crucial issue for new media scholars and practitioners at the start of the 21st century, as we experience another moment of technological and social shift to digital media. Radhika Gajjala, for instance, begins her feminist ethnography of a South Asian women's email list by asking: "Who speaks online, why, when, and how? Who counts as a netizen? What communities of practice and production shape and are in turn shaped by the Internet?" (2004, 2). These questions are especially pertinent to the 2008 web cartoon *Sita Sings the Blues*, which draws on the Orientalist image of the exotic, erotic woman in ways similar to the Fleischers and Ōfuji, but places it within new circuits of collaboration on the Internet.

Sita Sings the Blues is an 82-minute feature written, directed, computer animated and distributed entirely by Nina Paley, an independent Jewish film-maker from Urbana, Illinois. It is a musical comedy built around three parallel narratives. The first narrative is an animated adaptation of Valmiki's Sanskrit epic the *Ramayana*, as told from the perspective of Sita, the wife of Prince Rama of Ayodhya. Ever loyal, Sita goes into exile in the forest with Rama, where she is kidnapped by the demon Ravana. Rama, aided by the monkey god Hanuman, rescues his wife and regains his kingdom, but abandons her when his subjects criticize her for living under another man's roof (despite the fact that her purity was proven by a test of fire). The second narrative is an autobiographical story drawn from Paley's life, recounting how she moved to Kerala, India, with her husband, only to be dumped by him by email when she went back to New York on a short business trip. The third is a self-reflexive commentary track starring three shadow puppets voiced by friends of Nina's from Kerala, who improvise their own comically irreverent explanations of the *Ramayana*.

Paley uses a range of visual styles to tell this multilayered narrative, from "traditional Rajasthani miniature painting" (Chanda 2011, 6) to classic American graphic design, each corresponding to a different narrative thread. Significantly, Sita appears in most of the *Ramayana* song sequences with Betty Boop's signature circular, short-lashed eyes, oval facial structure and bow-lips (Figure 3). The resemblance is striking enough that it prompted film critic Roger Ebert to end a blog entry on the film with an image of a hat-tipping Betty captioned "Boo-boop a doop!" (Ebert 2008, n.p.). Paley's visual allusion is enhanced by the fact that each time Sita appears in this style, she "sings" a recording by 1920s jazz vocalist Annette Hanshaw, evoking Betty's jazzy performances and flapper style. The overall effect is a juxtaposition of temporally and spatially discrepant images even more striking than Ōfuji's straightforward recasting of Betty in a wholly Japanese setting. In a new era, and through new technologies, her approach to the exotic, erotic female singer is transformed. The task now is to discover to what degree these references retain the ambivalent yet persistent imperial complicities

Figure 3.   Sita's character design shows the ongoing influence of Betty Boop in *Sita Sings the Blues* (Creative Commons image).

Figure 4.   Lakshmi performs "Moanin' Low" to open *Sita Sings the Blues* (Creative Commons image).

of animation history, and to ask how the new medium may enable the creation of new images and social relations.

The most troubling aspect of shorts such as *Betty's Bamboo Isle* was the use of the "exotic erotic" ethnic stereotype. It may seem that Paley opens her film with the very same image. The establishing shot of her film fades in to show a screen full of sinuously patterned blue waves, over which synthesized sitar-like music plays. After a long moment, the tip of a crown emerges from the waves. It is the goddess Lakshmi, who will be incarnated as Sita. She rises from the water, curvy, clad in revealing pink, and literally sparkling with charm (Figure 4). At the flick of her hand, a peacock-gramophone rises from the waves so that Lakshmi can dance to the scratchy lines of a Hanshaw tune: "Moanin' low, my sweet man I love him so / Though he's mean as can be / He's the kinda man needs a kinda woman like me." At this point, however, the record begins to skip. The words "a woman like me" repeat until the goddess stops her dance in annoyance and lifts the stylus, sparking off a bang that leads into the film's energetic title sequence.

It would be easy to read this scene on the Betty Boop model of the exotic, erotic woman, as some critical online audience members have done. But on closer analysis, it seems that Paley's approach to this figure is more self-reflexive than the Fleischers'. While Betty Boop repeats her famous catchphrase during "live" performances in America and Japan, here the repetition is a skipping record, drawing attention to the artificial and "intermedial" (Chanda 2011, 3) quality of the performance. It is clearly mediated, not only through one technology, such as computer-generated imagery, but through multiple technologies, such as the gramophone recording, which interfere in the "natural" performance. The disruptive skip on the line "a woman like me" prompts us to ask: a woman like whom? A Hindu goddess? An American jazz singer? What are these women *like*, exactly? As in Judith Butler's analysis of the song "(You Make Me Feel Like) A Natural Woman", the fact that "woman" is defined by the simile "like" reveals

Figure 5.   Paley's version of Ravana in a World Laughter Day parade in Hyderabad, May 2009. Creative Commons image. Photo by Krishnendu Halder.

the constructed, performative aspects of gender (1990, 29–30). In emphasizing this line, Paley demonstrates a self-consciously parodic approach to the female role models provided by both nostalgic Americana and exotic Orientalism. Rather than simply "Americanizing" the *Ramayana* (as Ōfuji "Japanized" Betty Boop), she deliberately *plays through* exotic and erotic imagery. This creates what Graham Huggan, riffing on Spivak's "strategic essentialism", calls "strategic exoticism", in which "postcolonial writers/thinkers, working from within exoticist codes of representation, either manage to subvert those codes [...] or succeed in redeploying them for the purposes of uncovering differential relations of power" (2001, 32).

Parody and humour have played a large part in the subversive uses of "strategic exoticism" taken up by Paley's audiences in global audiences, who redeploy the film's exoticism in other contexts to draw attention to world issues. In May 2009, for instance, Paley's nine-headed, green-skinned version of Ravana showed up in a parade in Hyderabad celebrating World Laughter Day, a Mumbai-born social awareness event. He appeared as a large, brightly painted float bearing the word "RECESSION" in red across the teeth and phrases such as "GLOBAL SLOW DOWN" and "INFLATION" on the chest. This image (Figure 5) presents a complex negotiation between the "global" and the "local", the "exotic" and the "familiar". In using Ravana as an allegory for the multiple impacts of financial crisis, the float's creators in Hyderabad draw on a figure familiar across South Asia. But in modelling the float on Paley's American animation style, complete with English legends, they also recapture an air of the "exotic", presenting the familiar image of Ravana in a new way. Paley's Ravana is thus strategically redeployed through multiple understandings of what counts as "familiar" or "exotic", with the purpose of critiquing the economic inequality intensified by global recession (Figure 5).

And yet Huggan also cautions that "strategic exoticism" is not necessarily beyond neo-imperialism. Just as Ōfuji's parodic appropriation of Betty Boop drew on Japanese imperialist discourses of martial masculinity, so the postcolonial exotic is implicated in the dominant discourses of global capitalism in the 21st century. As Huggan notes, "the postcolonial exotic is, to some extent, a pathology of cultural representation under late capitalism – a result of the spiralling commodification of cultural difference" (2001, 33) seen in academic and cultural industries alike. Likewise, despite its initial online distribution, *Sita Sings the Blues* was not entirely outside the circuit of the art film industry. It fared well on the European film festival circuit, winning awards at the Berlinale and Annecy festivals, and garnered almost unanimous praise from mainstream reviewers such as Roger Ebert. As Huggan says of Salman Rushdie's playful mixes of kitsch Orientalism and western pop culture (2001, 69–76), there is an appealing cosmopolitanism to *Sita* that makes it highly consumable. Even the use of Paley's simplified Ravana in the World Laughter Day parade can be seen as part of a "celebration of hybridity" (Harindranath 2003, 157) that plays into some of the same logics of globalization (for instance, the use of English as a lingua franca) that it critiques.

Without forgetting the ever-present risk of complicity, however, I would still like to propose that Paley's new media distribution methods differ from those of established film markets. While material works (such as floats) must exist in one location and be made by a limited number of people, web animation offers new practical methods for the collaborative production, distribution and consumption of images. As a media activist, Paley has made every effort to avoid commodifying her film. During production, she continually sought reviews of half-finished scenes online, developing her work in direct response to commenters from around the world. She then released the entire film for free on her website under a Creative Commons Attribution-Share Alike licence, which allows audiences to download, share and remix the film at will. Since the film's release, she has worked to promote an animated "community of production and practice" based on the many-to-many structure of Internet distribution, disrupting the centre–periphery models of early film's imperial internationalism.

Unlike the 1930s, when distribution was largely centralized by western corporations with global branch offices, such as Paramount in Japan, Internet distribution involves a complex network of both corporate and grassroots participants. Those wishing to create a digitally animated film and distribute it online must still wrangle with corporations, as Paley found out in the expensive, difficult battle to license Hanshaw's music. But once that was done, Paley was able to distribute her film through decentralized and many-to-many forums that allow for different forms of engagement between herself and her audiences, including the creation of a "cultural commons" rather than a capitalist marketplace for her work. In an interview with Amy Chazkel for the journal *Radical History Review*, Paley (2011) addresses the ways in which a cultural commons may be created online, not through the restriction of licensed intellectual property, but through open-source distribution. Comparing the ownership of cultural properties to the seizure of land during the colonization of the United States, Paley (2011, 146) argues that new media need not follow colonizing capitalist logics. Instead, it may be based on the premise that freely sharing a media text does *not* dilute its value, as in commodity-based markets, but increases its effectiveness by creating more opportunities for viewers to participate in it. Voluntary donations and user-generated promotion through reposting, linking and blogging support this type of production. *Sita* was thus shaped not so much by the demands of capital, but by the "communities of production and practice" (Gajjala 2004, 2) that exist online.

Now, as Gajjala argues, it is important to ask *who* participates in such communities, and how they do it. The Internet, after all, does not exist in isolation from inequalities of access and ongoing conflicts in the material world. Much online engagement around *Sita Sings the Blues* took place on the message boards of blogs devoted to film reviews and cultural criticism, sites often lauded as democratizing tools for "radical technopolitics" (Kahn and Kellner 2005, 16). And yet these boards are not quite new "public spheres", in which everyone participates equally in rational discussion towards a common goal. In fact, online reactions to *Sita Sings the Blues* were violently divided along political and ideological lines in ways not seen in mainstream reviews like Ebert's, which were almost unanimously positive. On the one hand, the project received strong support from media activist NGOs such as Questioncopyright.org and from diasporic South Asian communities, such as the *Sepia Mutiny* and *Desifeminists'* blogs, who lauded it as a "feminist's retelling of Ramayana" that runs against misogynistic elements in the original work (Desifeminists 2009). On the other hand, it was also attacked by extreme right-wing Hindu nationalists in India and by western left-wing academics, both of whom, Paley claims, accused her of racism and "neocolonialism" (Di Justo 2008) just for adapting the *Ramayana* as an American. The most striking feature of these debates is not so much that they fall into "pro" and "con" camps along clear-cut ideological lines (though as I shall show, this also happened), but that they fractured along many intersecting and diverging lines of cleavage, including the issues of gender, ethnicity and nationality that have historically attended Betty Boop's travels.

In both polarized and more intersectional debates around this film, contention has arisen over the issues of cultural imperialism and appropriation many see in Paley's depiction of the exotic/erotic female body. One case where sharp divisions arose was the "Ban and Legal Action Against *Sita Sings the Blues*" petition, posted to the website ipetitions.com by an international coalition of right-wing Hindu organizations in March 2009. The petition writers particularly objected to the film's perceived disrespect in sexualizing religious figures, including "a semi-nude Goddess Lakshmi emerging from the sea and dancing on tune of a 'Blues' song [sic]" ("Petition" 2009, n.p.). Commenters in the United States, getting wind of the petition, then began to countersign with their own nationalist declarations, proclaiming "This is america [sic] we are free to comment and depict anything we want as long as it is not obscene" ("Petition" 2009, n.p.). Differing cultural standards as to what constitutes "obscenity" went unrecognized, and, in this particular instance, the petition led to a polarization of the debate into "religious fundamentalism" versus "freedom of speech" camps, effectively limiting true dialogue on the "exotic erotic" figure.

Still, there were also commenters in other forums who worked to raise self-awareness among the participants and generate more constructive debate. For instance, the second-generation Indian-American blog *Sepia Mutiny* hosted an interview between activist Tanzila "Taz" Ahmed and Paley in which Paley criticized the "Hindutvadi trolls" (Hindu nationalists who post with vitriol to provoke arguments) who accused her of appropriating their culture by making it about her own personal life. This post stirred up debate among commenters over "Who gets to say what is an acceptable appropriation and what is not?" ("Amardeep", commenting on Ahmed 2009, n.p.). Some attacked Paley's autobiographical approach, saying "This is an '*I'm pissed that my husband left me so I'll piss on the first husband in human history who left his wife*' kind of movie" (MoorNam). Others, however, took a more analytical response. One frequent commenter going under the handle "One Vaishnava's Opinion" posted 20 lengthy messages pointing to the many interpretations the Ramayana has elicited over the centuries. This poster essentially made the argument supported in Paula Richman's edited collections *Many Ramanyanas* (1992) and *Questioning Ramayanas* (2001): that there has been a long tradition of reinterpretation and cultural exchange around the *Ramayana*, including women's communal and personal interpretations of Sita's character in South Asia and abroad. Moreover, "One Vaishnava" stressed the necessity of confronting the issues raised by the film in the very forum of the online comments board, saying: "I believe that the Ramayana crosses and transcends time/ regions/cultures, but exactly *how it does that* – that is the discussion we are having!" (emphasis in original). In response, even hostile commenters grew more articulate in trying to establish requirements for respectful adaptations. In this case, we may see that online discussions of the film have enabled an explosion of contesting voices, sometimes positioned unequally with regards to language use and perceived authority to speak, but certainly not silenced.

The vocal debate between online commenters has in fact proved productive. As Paley explains in this very interview, because of her critics she became more aware of current South Asian politics and "honed [her] philosophies towards art and responsibility on the global stage" (Ahmed 2009, n.p.). This process exemplifies what ethnographer Anna Tsing calls friction: "the awkward, unequal, unstable and creative qualities of interconnection across difference" (2004, 4). In Tsing's model, "There is no reason to assume that collaborators share common goals. In transnational collaborations, overlapping but discrepant forms of cosmopolitanism may inform contributors, allowing them to converse – but across difference" (2004, 13). This is another way of working within and through viewpoints with which we may disagree, in a collaboration that includes productive contention.

Conclusion

Transnational collaborations through new media are not free of risks. As Gajjala says, new media technologies are still based on "sociocultural discourses and material practices that divide the privileged of the world from those less privileged" (2004, 106). Yet we must be careful not to mark the divisions too rigidly. In the cases of both *Betty Boop* and *Sita Sings the Blues*, I have shown how animators and viewers were in fact *multiply positioned* in their privileges and disadvantages. While the Fleischer Brothers worked from the minoritarian position of Jewish immigrants in the United States during the 1930s, they retained a level of privilege as male film-makers complicit in Orientalist conventions of representing race and gender. On the other hand, while Paley can be said to hold privilege as an American creator with access to technologies and educational opportunities denied to others, as a woman and a media activist she attempts to position herself differently in relation to her context, opening up a common space for collaboration online. In neither case are the centre–periphery binaries often seen as hallmarks of imperialism entirely applicable. But in neither case can we avoid the asymmetries of power that persist in new media. Rather, as "One Vaishnava" says, questions of *how* we participate in cultural exchange through media are what must concern us now. In order to avoid reducing the relations of animators and audiences to an easy polarization of dominance and resistance, I propose that we rethink new media production and reception at least partly beyond the old models of neo-imperialism.

Acknowledgments

For their support, I would like to thank the Social Sciences and Humanities Research Council of Canada, the University of Manitoba's Faculty of Arts, and the University of Stirling for hosting a wonderful PSA event. I would also like to extend my gratitude to Diana Brydon for her supportive comments on an earlier draft of this paper.

References

A Language All My Own. 1935. Dir. Dave Fleischer. Prod. Max Fleischer. Paramount.

Ahmed, Tanzila. 2009. "Sita Sings the Blues, Just for You." Sepia Mutiny. March 5. http://www.sepiamutiny.com/sepia/archives/005661.html.

Anderson, Joseph L., and Donald Richie. 1982. *The Japanese Film: Art and Industry*. Princeton, NJ: Princeton University Press.

Artz, Lee. 2003. "Globalization, Media Hegemony, and Social Class." In *The Globalization of Corporate Media Hegemony*, edited by Lee Artz and Yahya R. Kamalipour, 3–31. New York, NY: State University of New York Press.

Artz, Lee. 2005. "Monarchs, Monsters, and Multiculturalism: Disney's Menu for Global Hierarchy." In *Rethinking Disney: Private Control, Public Dimensions*, edited by Mike Budd and Max H. Kirsch, 75–98. Middletown, CT: Wesleyan University Press.

Betty's Bamboo Isle. 1932. Dir. Dave Fleischer. Prod. Max Fleischer. Paramount.

Bouldin, Joanna. 2001. "The Body, Animation and the Real: Race, Reality and the Rotoscope in Betty Boop." In *Affective Encounters: Rethinking Embodiment in Feminist Media Studies*, edited by Anu Koivunen and Susanna Paasonen, 48–54. Turku: University of Turku.

Butler, Judith. 1990. *Gender Trouble: Feminism and the Subversion of Identity*. New York, NY: Routledge.

Carlton, Donna. 2002. *Looking for Little Egypt*. Bloomington: IDD Books.

Chanda, Ipshita. 2011. "An Intermedial Reading of Paley's *Sita Sings the Blues*." *CLCWeb: Comparative Literature and Culture* 13 (3): 1–9. doi:10.7771/1481-4374.1798.

Chazkel, Amy. 2011. "Confronting the Enclosure of the Cultural Commons: An Interview with Nina Paley." *Radical History Review* 109: 137–152. doi:10.1215/01636545-2010-020.

Defeat of the Tengu. [Tengu Taiji, 1934.] 2008. Dir. Ofuji Noburo. Japanese Anime Classic Collection. Digital Meme.

Desifeminists. "Sita Sings the Blues." 2009. Desifeminists' Blog. Nov. 12. http://desifeminists. wordpress.com/2009/11/12/sita-sings-the-blues/.

Di Justo, Patrick. 2008. "One-Woman Pixar's Animated Film Premieres at Tribeca." *Wired*, April 25. www.wired.com/entertainment/hollywood/news/2008/04/sita?currentPage=all.

Dobbs, Mike. 2006. "Myron Waldman 1908-2006." *Cartoon Brew*, February 5. http://www. cartoonbrew.com/?s=Betty+Boop+%22Language+All+My+Own%22.

Dorfman, Ariel, and Armand Mattelart. 1975. *How to Read Donald Duck: Imperialist Ideology in the Disney Comic*. New York, NY: International General.

Ebert, Roger. 2008. "Having a Wonderful Time, Wish You Could Hear." *Roger Ebert's Journal*. Dec. 23. http://blogs.suntimes.com/ebert/2008/12/having_wonderful_time_wish_ you.html.

Gajjala, Radhika. 2004. *Cyber Selves: Feminist Ethnographies of South Asian Women*. Walnut Creek, CA: AltaMira Press.

Harindranath, Ramaswami. 2003. "Reviving 'Cultural Imperialism': International Audiences, Global Capitalism, and the Transnational Elite." In *Planet TV: A Global Television Reader*, edited by Lisa Parks and Shanti Kumar, 155–168. New York, NY: New York University Press.

Hendershot, Heather. 1995. "Secretary, Homemaker, and 'White' Woman: Industrial Censorship and Betty Boop's Shifting Design." *The Journal of Design History* 8 (2): 117–130. doi:10.1093/jdh/8.2.117.

Holberg, Amelia S. 1999. "Betty Boop: Yiddish Film Star." *American Jewish History* 87 (4): 291–312. doi:10.1353/ajh.1999.0035.

Huggan, Graham. 2001. *The Post-Colonial Exotic: Marketing the Margins*. London: Routledge.

Kahn, Richard, and Douglas Kellner. 2005. "Oppositional Politics and the Internet: A Critical/Reconstructive Approach." *Cultural Politics* 1 (1): 75–100. doi:10.2752/174321905778054926.

Klein, Norman M. 1993. *7 minutes*. London: Verso.

Krishnaswamy, Revathi. 2008. "Introduction: At the Crossroads of Postcolonial and Globalization Studies." In *The Postcolonial and the Global*, edited by Revathi Krishnaswamy and John C. Hawley, 2–21. Minneapolis, MN: University of Minneapolis Press.

Mattelart, Armand. 2003. *The Information Society: An Introduction*. London: Sage.

Richman, Paula. 1992. *Many Ramayaṇas: The Diversity of a Narrative Tradition in South Asia*. New Delhi: Oxford University Press.

Richman, Paula, ed. 2001. *Questioning Ramayanas: A South Asian Tradition*. Berkeley: University of California Press.

Sita Sings the Blues. 2008. Dir. Nina Paley. http://www.sitasingstheblues.com/.

Standish, Isolde. 2006. *A New History of Japanese Cinema: a Century of Narrative Film*. New York, NY: Continuum.

The Motion Picture Production Code of 1930 (Hays Code). 2006. ArtsReformation.com. 12 April. http://www.artsreformation.com/a001/hays-code.html.

"The Petition." 2009. Ban and Legal Action Against Sita Sings the Blues. http://www.ipetitions.com/petition/sitasingstheblues/.

Thompson, Kristina. 1985. *Exporting Entertainment: America in the World Film Market 1907–1934*. London: BFI.

Tomlinson, John. 1991. *Cultural Imperialism*. Baltimore, MD: Johns Hopkins University Press.

Tsing, Anna Lowenhaupt. 2004. *Friction: An Ethnography of Global Connection*. Princeton: Princeton University Press.

Tsutsui Yasutaka. 1992. *The Legend of Betty Boop: the Actress as Symbol, the Symbol as Actress*. [*Beti Būpu Den Joyū to Shite no Shōchō Shōchō to Shite no Joyū*] Tokyo: Chuokoran.

Yamaguchi, Katsunori and Watanabe Yasushi. 1977. *History of Japanese Film Animation*. [*Nihon animēshon eigashi*] Osaka: Yūbunsha.

Pluralism and cultural imperialism in the network films *Babel* and *Lantana*

Vivien Silvey

Australian National University, Canberra, Australia

The multinational film *Babel* (2006) and the Australian film *Lantana* (2001) dwell on controversies surrounding international and multicultural relationships. They belong to a genre termed network cinema, which focuses on social problems and responds to the paradigm of network society. *Babel* and *Lantana* aim to represent cultural pluralism, yet their narrative politics compromise this aim. Although they seek to decolonize the marginalized and give equal voice to broad cross sections of society, these two films rely on narrative tactics which colonize, stereotype and essentialize cultural others. This paper argues that the ways in which the two texts reterritorialize multicultural and gendered relationships undermine their thematic concern with the need for cross cultural understanding and trust. In many ways these narrative tactics align with those popular in Hollywood films, which challenges the notion that world cinema consists of oppositional film industries. Nevertheless, the production and distribution of these two films reveal concretely asymmetrical power relationships. These findings suggest that the common notion of a clear divide between Hollywood and national art cinemas (implicit in the term "world cinema") requires revision.

Babel (Iñárritu 2006) and *Lantana* (Lawrence 2001) belong to a genre described as network cinema that has flourished around the world since the early 1990s. Well known examples include *Code Unknown* (Haneke 2000), *Crash* (Haggis 2004), *Magnolia* (Anderson 1999) and *The Edge of Heaven* (Akin 2007). This genre belongs to a broader contemporary imaginary concerning social networks. For instance, David Mitchell's novel *Cloud Atlas* (2004) and Sebastian Faulks's *A Week in December* (2011) similarly depict networks of interconnecting strangers and acquaintances. While there is contention over how to label the film genre (see Quart 2005; Everett 2005, 159; del Mar Azcona 2010, 1; Pisters 2011), I take as my point of departure the point made by both David Bordwell and Paul Kerr that "network cinema" is the most appropriate term (Bordwell 2007, 191; Kerr 2010, 40). It signals both the films' social thematics regarding the paradigm of network society (Castells 2000, 3) and industrial practices (since the films frequently rely on networked modes of production and distribution). As well as networked production technologies, home-viewing technologies have been instrumental in the genre's popularity. DVDs and the Internet have provided both artistic and commercial incentives for film-makers to convey complex narratives which require multiple viewings to solve the narrative puzzles (Bordwell 2006, 74).

Network films characteristically show multiple protagonists who occupy individual narrative threads. These characters share similar thematic and emotional concerns, although they are often strangers to one another. Frequently these central characters represent contrasting cultural identities, and the films concentrate on crossing borders, experiences of marginality and multiculturalism. The genre thematically represents the world as pluralistic, yet the viewer is also prompted to conceive of the characters as a cosmopolitan community. This conception of global and multicultural networked communities importantly contrasts with the common association of community with spatial, familial, cultural and racial bounds.

Since network films have appeared around the world and are often products of international collaborations, they suggest that the term "world cinema", with its implication of a dichotomy between popular Hollywood and other national cinemas, is outdated. World cinema commonly implies art cinema from nations outside of Hollywood. Hollywood is typically defined as commercial entertainment cinema espousing Americentric patriarchal ideologies and is distributed widely and successfully throughout much of the world (Stam and Spence 1983, 6). In contrast, films from other nations which are distributed through international film festivals and art house cinemas are commonly venerated as art cinema. Art cinema's attributes are said to include realism, open endings, ambiguity, intellectual rigour and a focus on characters who are culturally marginalised (Bordwell [1979] 2002, 95–99; Neale 1981, 13–14). Recently, major Hollywood studios have teamed up with independent studios to make commercially successful yet artistic films which screen at festivals, occupying a middle ground between the Hollywood versus art cinema dichotomy (Scott 2004, 35; Cooke 2007b, 3). To some, this represents an imperialistic exploitation on Hollywood's part (Naficy 2008; Cooke 2007b, 3).

Currently, the term world cinema is under academic revision. Lúcia Nagib proposes that world cinema be thought of as "a positive, inclusive, democratic concept [...] [that] allows all sorts of theoretical approaches, provided they are not based on the binary perspective" of Hollywood versus its others. This is an approach which takes into account how films circulate around the world as "a global process" (2006, 35). Nagib argues that comparative analyses of world cinema be attentive to cross-cultural, networked relationships and the power dynamics involved in these relationships. Paul Willemen also advocates comparative studies in world cinema in order "to find ways of overcoming the limits of any cultural relativism, any fetishization of geo-political boundaries, and to elaborate a cultural theory worthy of the name" (2005, 98). I admire these new takes on the term world cinema for their opposition to the implicitly elitist binary implications of the term. I believe that comparative analyses can reveal some of the ways in which this dichotomy is problematically simplistic.

Babel and *Lantana* provide a comparative axis which cuts across the binary of Hollywood and national art cinemas. *Babel* and *Lantana* represent very different industrial and cultural backgrounds. The former is an international co-production that has major Hollywood financing and contrasts First World and Third World characters. The latter is an Australian film that focuses on the tensions of multiculturalism inherited from the country's settler identity. As multinational and national examples of the network genre, the two films problematize the notion that world cinema consists of exclusive categories. For instance, Marina Hassapopoulou (2008) gives an exceptionally nuanced approach to the issue, attending to *Babel*'s use of conflicting art and mainstream cinematic styles. Yet *Babel* and *Lantana* are frequently classified in relation to this binary, which suggests that the categories of national cinema and Hollywood retain their significance. In his illuminating essay "*Babel*'s Network Narrative: Packaging a Globalized Art Cinema", Kerr (2010, 48–49) suggests that *Babel*'s production and distribution processes indicate a strong degree of Hollywood imperialism. Naficy also describes *Babel* as an example of "Hollywood […] chang[ing] in order to remain the same" (qtd. in Silvey 2009, n. pag). Studies of *Lantana* such as those by Duncanson, Elder, and Murray (2004) focus on its tropes as national cinema. I contend that the films challenge the binary of the term "world cinema", although they are simultaneously symptomatic of continuing asymmetry between Hollywood's hegemony and the marginalization of other cinemas. This article focuses on comparing the narrative and thematic aspects of the two texts before moving on to explore their production backgrounds.

Written by Guillermo Arriaga and directed by Alejandro Gonzalez Iñárritu, *Babel* follows four narrative threads. It tells the story of two boys, Ahmed and Yussef, in the mountains of Morocco. Their father Abdullah buys a gun from a man named Hassan and gives it to the boys to keep jackals away from their goats. They test its range by shooting at a bus in the distance, and Yussef's bullet hits the target. Inside the bus, the American tourist Susan (Cate Blanchett) is hit in the shoulder. The Moroccan tour guide Anwar takes Susan and her husband Richard (Brad Pitt) to his village Tazzarine, where the local vet sews her wound without anaesthetic. Richard tries to get help from the American Embassy, where staff delay their assistance because they perceive the shooting as a terrorist act. Meanwhile, Richard and Susan's two children, Mike and Debbie, are in America under the care of their Mexican nanny Amelia (Adriana Barraza). Because of the film's jumbled plot, we see Amelia take Mike and Debbie to her son's wedding in Mexico in scenes cross-cut as if simultaneously occurring with the film's other three threads. After the wedding, Amelia's nephew Santiago (Gael Garcia Bernal) flees from American border guards and in the process abandons Amelia, Mike, and Debbie in the desert. In the linear chain of events, this episode occurs after Richard and Susan arrive safely at a hospital in Casablanca. The fourth thread, set in Japan, concerns a deaf Japanese girl named Chieko (Rinko Kikuchi) whose sexual encounters reveal her private struggles with social belonging, self-esteem, and grief for her mother's suicide. Her link to the other threads is through her father Yasujiro (Kôji Yakusho). Yasujiro had been on a hunting trip to Morocco and had given the gun as a gift to his guide Hassan. As the stories proceed, Mike, Debbie and Amelia are rescued from the desert, and Amelia is deported. Richard and Susan recover safely from their ordeal and return to America. The police hunt down Yussef, Ahmed and their father. In a shoot-out, Ahmed is killed and Yussef and his father are arrested. The final scenes show that Chieko finds platonic comfort in a police officer's sympathy and she reconciles with her father.

Directed by Ray Lawrence, *Lantana* is an Australian film set in Sydney. The film opens with a shot of a woman's corpse tangled in the lantana plant's undergrowth. A species introduced from South America, lantana is regarded as a weed in Australia. The movie revolves around ten characters. Valerie (Barbara Hershey) is an American psychologist, living in Australia with her husband John (Geoffrey Rush), a Dean of Law. Their marriage has suffered since the murder of their daughter 18 months before. John is increasingly withdrawn and Valerie begins to suspect he is having an affair with one of her patients, a gay man named Patrick (Peter Phelps) who is seeing a married man. Sonja (Kerry Armstrong), another of Valerie's patients, is married to a police officer named Leon (played by Anthony LaPaglia, who is of Italian and Dutch heritage and known for his American film roles). Leon has cheated on Sonja with Jane (Rachael Blake), a woman from their Latin dance class who is separated from her husband Pete

(Glenn Robbins). Ironically, Leon happens to meet and drink with Pete one evening after Pete bumps into Valerie on the street. Jane is a neighbour of Nic and Paula, a working-class couple with three young children. Nic is played by Vince Colosimo, known for his Greek heritage, and Paula is played by the Italian Australian actress Daniela Farinacci. As the film unfolds, Valerie's car breaks down one night and, after leaving an answering machine message telling John she still loves him, she catches a lift from a stranger. When she has not returned home the next day, Leon and his fellow detective Claudia, an Aboriginal woman played by Leah Purcell, investigate Valerie's disappearance. Leon meets the characters Valerie knew and discovers that Sonja suspects his own infidelity. Jane reports Nic to the police when she sees him throw a woman's shoe into the bushes. We discover it was Nic who gave Valerie the lift, but her death was accidental. She had leapt from the car into the bushes, having misinterpreted Nic's intentions when he took a shortcut. The film ends with a montage of the characters in various states of loneliness and uncertainty. Jane and Pete remain separated, Nic and Paula no longer respect Jane, Patrick has little chance of a future with his lover, and John is burdened by grief. The only hopeful ending is Claudia's; she finally has dinner with a man she has a crush on, who happened to bump into Leon one day while jogging. The last shot lingers on Sonja and Leon dancing slowly, unsure of the future of their marriage.

Network films and the narration of pluralism

Babel and *Lantana*'s character networks aim to re-imagine pluralistically societies and communities that are usually conveyed as distinct and separate. Their collections of characters cross cultural boundaries and express marginal voices in relation and in conflict to dominant ones. In response to global inequalities, *Babel* conveys a cosmopolitan theme of international human connectedness. The scenes set in Morocco and Mexico allude to America's response to 11 September 2001 and America's controversial border politics with Mexico. The film represents the disjunctive experiences of wealthy westerners from the US, Europe and Japan being able to tour in Morocco, while the Moroccan characters (who live in poverty compared to the westerners) are restricted under the influence of US-dominated global politics. The Mexican characters are likewise economically disadvantaged and are intimidated by the Americans. In each scenario, characters similarly experience suffering, social marginalization and the inability or struggle to communicate with authorities. *Babel* creates sympathy for each of the characters, placing its Hollywood stars on the same stage as the non-US actors. The film therefore appears to counter dominant narratives which demonize and/or marginalize such figures and insists that communication yields a solution to these problems (Shaheen 2003). It condemns miscommunication with the thematic implication that, as one tag line states, "tragedy is universal". Through this condemnation, *Babel* links together different people and societies in a bid to unsettle and reassess Americentric perspectives.

In the guise of a thriller, *Lantana* observes a social cross section of Sydney residents, intimating the controversial nature of Australian multiculturalism. Instead of tokenistically foregrounding characters' racial and ethnic identities (which are marked by the choice of actors), *Lantana* presents a society in which this issue is ignored. Nevertheless, the film uses a pluralistic form as it follows different characters, in separate threads, who articulate disagreements and dissatisfactions with one another. For example, in Valerie's counselling sessions we see Sonya's worry about her marriage, which she does not admit to Leon, and Valerie's own unspoken worry that her husband is having an affair. We also see Patrick's hostility to Valerie because she is uncomfortable with his participation in an affair, but later we are also driven to sympathize with Patrick. On a racialized level, when the Aboriginal police officer Claudia reprimands Leon for his brutality and marital indiscretions, this presents her in a position of authority, in contrast to Australian narrative traditions which privilege white Australians wielding power over Aboriginal people (Duncanson 2009, 45–46). The film's pluralistic form and concentration on mistrust portrays Australian multiculturalism as troubled. By the end of the film, the dispersal of characters and the rifts between them suggest that a tolerant, multicultural Australia has failed. For instance, the second- or third-generation Italian immigrant Paula forbids the white Australian Jane from coming near her house and family again, the gay man Patrick is abandoned by his lover, and the white Australian John looks out mournfully at the bush so commonly associated with Aboriginal land rights, but here (due to the imported lantana weed) associated with colonial violence committed on the land. The uncertainty of Leon and Sonya's future together symbolizes an overarching uncertainty of harmonious and trusting relationships. Duncanson et al. compellingly argue that throughout the film the idea of Australian "neighbourliness" and mateship breaks down (just as marriages and masculinity break down), exposing the resentment and boundaries that simmer under the rhetorics of the "good neighbour" and tolerant multiculturalism (2004, 12–15). Thus the supposedly community-forming bases of trust, loyalty, settlement, neighbourliness, gender identities and the nuclear family are each unsettled in *Lantana* to show the shaky grounds on which Australian multicultural harmony is constructed. In this way, I believe, the film represents a valuable critique of the ways in which Australia's governments and citizens have responded to xenophobia and gender inequalities.

While *Babel* and *Lantana* offer pluralistic narrative formats and represent multicultural and international communities, their constructions of power relationships deserve close attention. The fact that these two films from such different backgrounds follow comparable trajectories raises crucial questions about whether the Hollywood versus national art cinema dichotomy is useful as a marker of contrasting narrative ideologies. Their similarities suggest that although *Babel* is often regarded as a Hollywood film and although *Lantana* is regarded as an art house film, their narrative differences are not as distinct as would typically be thought.

Traditionally, Hollywood films marginalize and stereotype women by perpetuating patriarchal ideologies (Kaplan [1983] 1988, 5). We might assume by default of their oppositional reputation that non-Hollywood cinemas feature women in prominent nonromantic roles and critique patriarchal norms. Yet, indicative of the fact that art cinema is often unduly considered exempt from the same types of criticism directed at Hollywood, more attention has been paid to gender representations in Hollywood than in art cinema (Devereaux 1990, 340; O'Regan 1996, 334). Key to both *Babel* and *Lantana* is the figure of a bereft American mother wounded in a foreign country. The victimization of Valerie and Susan falls back on tropes of patriarchal and colonizing power relationships, reinstating Leon and Richard as the films' central characters. Both women initially demand "too much" of their husbands and are subsequently relegated into positions of passivity, which restores patriarchal order (Wood 2007, 148). The early scenes of *Babel* illustrate Richard and Susan's tense marriage and demean Susan as a paranoid tourist. She rudely throws out the ice in her glass at the café, believing it to be contaminated, whereas Richard tries in a hushed tone to temper her rudeness. Susan's agitation in this scene stands in stark contrast to Richard's weariness, suggesting that although Richard abandoned Susan after the death of their infant, Susan's continued anger in spite of Richard's apology is an unfairly held grudge. When Susan is wounded, Richard takes charge as a loyal husband, yelling at people in order to get results and caring tenderly for Susan. It is only when Richard thus becomes the active and heroic husband and Susan is victimized, weakened and dependent on him that they reconnect. Tellingly, despite his deliberately aged and "unglamorous" appearance, Brad Pitt is filmed in ways that underscore his stardom and locate him as the central character (Hassapopoulou 2008, 2). As Hassapopoulou suggests, "the camera still lingers on his face on numerous occasions, reminding us of his internationally recognizable face" (2008, 2). Structurally, these American characters, especially Richard, take the spotlight. Richard and Susan's experiences affect all of the other narrative threads, and it is only through Richard and Susan that Amelia's and Yussef's stories are linked to each other. Whereas Yussef has no verbal contact with Susan, Richard speaks with Amelia over the phone. *Babel* therefore places the two internationally recognizable anglophone stars at the heart of its jumbled narrative, from which the American man emerges as the hero who protects the woman from the dangers of a foreign country. Arguably, *Babel* inscribes similar tropes of patriarchal and cultural imperialism closely associated with Hollywood narratives, for example *King Kong* (Cooper and Schoedsack 1933), *Independence Day* (Emmerich 1996) and *The Omega Man* (Sagal 1971; Davies 2005; Miller 2005).

In *Lantana*, Valerie and John's marriage is similarly strained. They grieve for the murder of their daughter Eleanor, and Valerie does not trust John's fidelity. Valerie expresses her grief by writing a book about and maintaining a shrine for Eleanor, which is presented early on as an attention-seeking "indulgence" (Wood 2007, 148). In contrast, it is later revealed that John privately visits the site of Eleanor's murder, legitimizing his grief as personal rather than exhibitionistic (148). Valerie's disappearance acts as a catalyst for the resolution of male characters' emotional crises. The investigation allows John to confess his emotions about his marriage and infidelity to Leon. His stoic honesty (accompanied by a glass of whiskey) recapitulates his masculinity. The investigation similarly prompts Leon to confront his own marital problems. In both films these representations normalize uneven gender power relationships, objectifying and portraying women as hysterical and demanding. The narratives pacify their voices while valorizing masculine norms of strength, action and endurance. The films therefore challenge the understanding that Hollywood and non-Hollywood films are ideologically opposed in their representations of women.

Babel and *Lantana* recall non-mainstream films which provide character depth to non-heterosexual characters and to characters who are identified in part by their sexual behaviour. *Lantana* counters the stereotype of effeminate gay men with Patrick's characterization as a critical, somewhat unkind man who is angered by homophobia. Yet Patrick's aggressiveness towards Valerie and his obstinacy towards Leon are punished with a narrative resolution which shows him standing in the rain miserably watching his lover return to his wife and children. *Babel* destabilizes the stereotype of the sexualized Japanese schoolgirl by showing Chieko's use of this identity as an attempt to combat her depression. Curiously, Iñárritu states: "Deaf girls have a great sexual energy, I don't know why. They think about it all the time." The acting coach adds: "I wouldn't say deaf equals erotic, but [Chieko] can easily be seen that way" (Iñárritu 2006, DVD featurette). These points of view intersect with both the practice of eroticizing schoolgirls in Japanese visual culture and the Orientalist stereotype of quiet, young Asian women. Yet Chieko's sexual advances contravene the passivity inscribed in the stereotypes and expose her emotional desperation and anger at being marginalized. Nevertheless, Chieko is inscribed as a victim, whose sexuality threatens others and who needs to find comfort in patriarchal figures in order to feel socially accepted and recover from the trauma of her mother's suicide. Thus, alongside its objectification of Susan and Amelia, *Babel*'s portrait of Chieko situates the female body as a site of suffering and hysteria in need of the restoration of patriarchal order (Wood 2007, 149; Hassapopoulou 2008). Both *Lantana* and *Babel* therefore restore patriarchal order in response to their sexually transgressive characters.

Hollywood is often seen as demonizing, stereotyping and marginalizing cultural others, whereas national art house films are thought to question these practices and privilege marginalized people in central roles (Stam and Spence 1983, 8). *Babel* and *Lantana* challenge this apparent opposition. In both films, cross-cultural relationships exhibit problematic gendered and racialized power relationships. Duncanson argues that Claudia represents "a quiet suggestion of a more legitimate legal authority" than Leon or John's failed attempts to find a criminal or legal order (2009, 34). While this point is true, Claudia's peripheral narrative position is troubling. We receive little insight into her life beyond Leon's company, and Leon oversees her romantic life, since he asks her about it and offers her advice. In turn, Claudia is morally compromised when she lies to Sonya about Leon's infidelity. Whereas Claudia occupies the margins of the narrative, Leon is a central character. *Lantana* therefore marginalizes the Aboriginal other while centralizing the Italian-white Australian character.

Similarly, in *Babel* cosmopolitan relationships between western and non-anglophone characters are problematic. Richard and Anwar's friendship portrays Anwar as a servant to Richard. He never expresses anger towards Richard, even when Richard yells and swears at him and other people. We do not learn anything about Anwar's life, outside of keeping Richard company and helping him. When leaving the village, Richard attempts to pay Anwar, but Anwar refuses the money. This scene shows the critical disjunction between the American's belittling attitude that money can repay kindness and the Moroccan's altruistic desire to help. Nevertheless, this motif underpins Anwar as a humble servant who has taught Richard the meaning of generosity and kindness to others. On a related note, in Tazzarine the wizened old woman who lets Susan smoke her intoxicating pipe is presented exotically. The camera wavers over her face in close-up from low angles, framing her as mysterious. Her smoking recalls orientalist tales of the other's magical healing powers (one reviewer notes that this misrepresents Moroccan customs; see Mezgarne 2007). In their figuring of cosmopolitan relationships between members of western and non-western cultures, both *Lantana* and *Babel* effectively marginalize the latter voices.

In showing rifts between anglophone westerners and their cultural others, the two films' portraits of cultural others further display the trope of othering. *Babel*'s jumbled plot throws viewers into a state of confusion in order to emphasize the characters' disjunctive experiences, but it structures the narrative in ways that create bias against the non-anglophone characters. Kerr argues that *Babel*'s plot "functions to conceal the irresponsibility of Amelia in opting to take the two children in her charge to Mexico for her son's wedding, despite her awareness that their mother has been shot" (2010, 42). The film places Amelia's end of the phone call, in which Richard tells Amelia that he cannot find a babysitter because Susan has been shot, at the beginning of the film. We are encouraged to sympathize with Amelia's desperation to get to her son's wedding because, from what we can hear, at this point it seems simply that Richard failed to arrange a babysitter as he had promised. Thematically this scene critiques American employers placing unfair demands on foreign illegal workers. Yet by the end of the film the narrative reveals Amelia's actions to be irresponsible rather than justified. The border officer who interviews Amelia and suggests she take voluntary deportation is not portrayed as villainous and cruel, but as a matter-of-fact blue-collar worker. He reminds Amelia that she should have thought before committing her actions. This remark, and the fact that a few scenes later we see Richard's side of the phone call, in which he bravely stifles his tears so as not to upset his son, highlights the gravity of Richard and Susan's situation and underscores Amelia's irresponsibility. Whereas the film presents Richard and Susan as undeserving victims, Amelia's fate is retrospectively framed as something she has brought on herself.

Amelia's nephew Santiago is similarly presented as an irresponsible Mexican. Kerr succinctly notes that "whilst the multi-character, multilingual structure appears to present an equitable image of a dislocated world it is only, in fact, the Mexican and Moroccan characters who are shown to use weapons irresponsibly" (2010, 42). Illustrating this point, Amelia's nephew Santiago behaves in ways that actualize the terms of his early joke describing Mexicans as dangerous. He drives while drunk, unwisely provokes the border guard and drives recklessly into the desert. His foolish actions lead to Amelia and the children's near deaths and Amelia's deportation. The film skirts around the fact that Mexican immigrants die trying to get into America because they are seeking to escape poverty. Although we get a glimpse of people who have experienced such desperate conditions, imprisoned in the back of the patrol van that picks up Amelia, Santiago's dramatic and reckless flight into the desert obscures the real tragedy of border issues. Santiago's characterization provides an essentialist portrait of Mexicans as reckless, anti-authoritarian and stupid (Orrenius 2001, 9), just as Amelia's attempt at this border crossing is branded as foolish; both characterizations reiterate an evasive approach to the representation of pertinent border issues.

Babel characterizes the Moroccan boys Yussef and Ahmed as foolish and naive. The fact that Yussef and his sister harbour an incestuous attraction to each other perpetuates stereotypes of Third World children as uncivilized and ignorant of social taboos. The use of handheld camera, location shooting and the use of local untrained actors for the roles of the boys is supposed to underscore that the thread strongly resembles reality. In drawing an analogy to the Bush Administration's response to 11 September 2001, the film conflates Morocco with the Middle East. This tactic avoids any direct and specific address of the topic. By showing a perceived act of terrorism as an unintended act of foolishness, *Babel* avoids considering reasons why the terrorist events of the early 2000s occurred. When Yussef gives himself up to the police because he feels responsible for Ahmed's death, it normalizes the representation of the cultural other as repentant for his faults. Similar to Amelia's response to Mike when she tells him she is not a bad person but has done something stupid, Yussef becomes aware of the consequences of his actions and his mounting fear and guilt reaches critical mass. His surrender comes as a narrative catharsis which restores law and order, although his family is destroyed. Amelia's and Yussef's threads both result in the law reining in these irresponsible characters, as they both acknowledge and repent their guilt and foolishness, thereby affording narrative catharses. These instances naturalize the characters as simply foolish and repentant, sidestepping direct issues of hostility between America and the Middle East or Mexico.

Lantana similarly frames the working-class Greek Australian Nic as an irresponsible other. Recalling Santiago in *Babel*, Nic drives while drunk and offers Valerie a lift. His decision to take a shortcut without explaining himself prompts Valerie to jump out of the car and fall to her death. While Nic is proven not guilty, the scene which shows him crying in Paula's arms reiterates his emasculating portrait as an irresponsible and unemployed stay-at-home father whose wife treats him like a child (Wood 2007, 148). The final scenes portraying Nic and Paula happily resting on their lawn with their children appear to idealize them as the perfect trusting family. Yet it frames them from a distance, reinstating a stereotype of the poor yet happy and innocent migrant other. In this respect, both films use narrative devices of plot, framing and characterization to perpetuate othering perspectives of marginalized people.

Hollywood and non-Hollywood films are often distinguished by the degree of narrative closure and their thematic concerns (Bordwell [1979] 2002 96–97; Everett 2005, 164). Traditionally, Hollywood errs towards conclusive denouements with conservative themes of restored social and familial harmony. Non-Hollywood films typically choose more ambiguous and unresolved endings (Everett 2005, 163–164). Both *Lantana* and *Babel*'s final scenes undermine the complexity of the issues of gender politics, multiculturalism and global inequalities that they have raised. They have been criticized for reinstating norms of white patriarchal power (Duncanson, Elder, and Murray 2004, 3; Hassapopoulou 2008). *Babel*'s final slow zoom out from Chieko and her father's embrace to show Tokyo's night lights emphasizes the theme of connectedness. Its implications of omniscience and closure subsume the preceding traumas. *Lantana*'s final montage situates its characters in positions which "[produce] a vision of subtly normalised hetero, mono, familial" relationships (Duncanson, Elder, and Murray 2004, 3). Both *Babel* and *Lantana* marginalize cultural others and reinstate Eurocentric and Americentric patriarchal and imperialist orders. This is in spite of their attempts to "[make] us feel and experience that this [global inequality] is an affair of everyone" (Pisters 2011), to represent forms of cultural pluralism and to create character depth which challenges emotionally distancing racial stereotyping. Therefore, although both films convey themes and characters associated with art cinema, they use representational politics typically associated with Hollywood cinema. In addition to these narrative trajectories, the films' networks of production and distribution deserve attention since they are key influences on how the films are categorized within the limits of Hollywood and art cinema.

Babel and *Lantana's* production and distribution: imperializing practices in action?

The rise of co-productions has prompted discussion of the ways in which international financial and cultural collaborations facilitate the products' international distribution and potentially challenge the dominance of Hollywood's distribution around the world (Dennison and Lim 2006; Cooke 2007a; Halle 2008, 6). Such "global process[es]" help to break down perceptions of the insularity of national cinemas and to embrace a more fluid, cross-cultural and dialogic perspective (Nagib 2006, 35). For example, attention is being paid to the fact that cinemas of India and East Asia dominate their local and related markets in counter to the long-held claim that Hollywood has a "global" hegemony. Nonetheless, these industries' international reaches remain disproportionate to that of Hollywood (Higbee and Lim 2010, 15–16; Virdi 2003, 21–22). Many network films, including *Babel* and *Lantana*, are multinational co-productions, which suggests that they may challenge Hollywood's dominance over global distribution. However, an investigation into their production and distribution illustrates that unequal power structures between particular cinemas remain in place.

Lantana was distributed through film festival channels, which tend to emphasize national specificity (Chan 2011, 255). Brian MacFarlane (2002, 160) argues that it occupies a niche market, which embodies Australian cinema's only hope of countering Hollywood's dominating presence. *Lantana* was received well at international festivals including Toronto and Athens, and Telluride in the US. In Australia it gained widespread recognition and ran for extended seasons at multiplex cinemas (Collins and Davis 2004, 34). The box office takings reflect its moderate success as a national art house Australian film, doing well in comparison to other non-US art house films (Box Office Mojo 2011; IMDb 2011a). *Lantana*'s production and distribution processes strongly identify it as an art house Australian film, despite its international elements.

Lantana gained most of its Australian funding from Jan Chapman production studios and the government funding bodies of the New South Wales Film and Television Office and the Australian Film Finance Corporation. It was in part produced by a small investor-run German company MBP, based in Munich (MBP 2011). Aside from this German financial input, *Lantana*'s casting points to international appeal. Indicative of the trend of "Australians in Hollywood", Geoffrey Rush is known for his roles in Hollywood films as well as Australian films. Anthony LaPaglia is similarly known for his presence in US movies and television shows (Ebert 2002). The presence of American actress Barbara Hershey in *Lantana* is emphasized as foreign and underscores the film's theme of colonialism in Australia. This is similar to Laura Linney's role in the director's subsequent film *Jindabyne* (Lawrence 2006), and represents a trend of Australian co-productions starring non-Australians, including *The Boys Are Back* (Hicks 2009, starring Clive Owen) and *December Boys* (Hardy 2007, starring Daniel Radcliffe). Despite these foreign inputs, *Lantana* is generally identified as a "a relatively low budget" national film (Film Finance Corporation Australia 2003, 7) and its use of familiar Australian actors and locations grounds these motifs of glocalization as emblems of Australia's national identity.

Like *Lantana*, *Babel* was first released on the film festival circuit, yet it gained a far broader global distribution than *Lantana*. It showed at Cannes and at least 19 other festivals around the world, including Marrakech, throughout Europe, Indonesia and Brazil. It also gained an international general release across 42 countries (IMDb 2011b). The fact that it was written and directed by Guillermo Arriaga and Alejandro Gonzalez Iñárritu, and that it had festival releases which conferred numerous nominations and awards, mark out *Babel* as an independent film which evolved into an international blockbuster. Its star cast included Mexico's rising star Gael Garcia Bernal and the well-known Japanese actor Kôji Yakusho in supporting roles. The film also featured upcoming talent Rinko Kikuchi and Mexican television soap opera star Adriana Barraza in central roles. This multinational collection of stars represents an international scope indicative of co-productions. In spite of this, *Babel*'s Hollywood associations raise questions about its status as an art house film. Paul Kerr (2010, 46) suggests that its international elements were included in part to capitalize on their appeal to global markets, which suggests that the film indeed imperialistically enlisted "foreign" actors to earn Hollywood a global profit. *Babel*'s narrative focus on A-list stars who are simultaneously known as stars of independent cinema and outsiders from the Hollywood mainstream (Brad Pitt and Cate Blanchett) relies on their stardom to attract audiences. This is evidenced by the promotional posters on which these Hollywood stars are given bigger facial close-ups than Rinko Kikuchi, Gael Garcia Bernal or Adriana Barraza. In contrast, the casting of Moroccan non-actors relies on essentialist tropes, since local non-actors were hired. This tactic has resulted in a lack of consensus by bloggers and commentators as to whether the film fits into Hollywood or art house categories (Frazer 2006; yndprod-2 2006). Indicating a strong indictment of the film's Hollywood background, the film's Americentrism has been debated in several blogs which question the authenticity of the languages and locations used (AllDeaf.com 2007; Mezgarne 2007).

Just as *Babel* narratively expounds Americentric viewpoints despite its multinational focus, its funding backgrounds and geographies of distribution imply an American-driven colonization of typically non-Hollywood territory. The ambiguity of Hollywood's changing identity is reflected in studios such as Fox Searchlight, as they invest in independent productions (Scott 2004, 35). Kerr's detailed description of the funding bodies contributing to *Babel*'s production and distribution indicates that the input of the "American companies, mak[e] it, in financial terms at least, a predominantly American film" (Kerr 2010, 45–46). *Babel*'s distribution circuit therefore illustrates a trend in which American films backed by companies closely related to Hollywood studios saturate international film festivals and go on to general releases internationally (Cooke 2007a, 3). *Babel*'s budget was 25 million US dollars, a large amount given to director and writer Iñárritu and Arriaga because of their earlier success with *21 Grams* (2003) and *Amores Perros* (2000). In contrast to *Lantana*'s small budget and predominantly localized national identity, *Babel* was produced and distributed with a much broader reach, thanks to its Hollywood funding. *Babel*'s six nominations and one win (for music) at the Academy Awards legitimate its mainstream recognition. Iñárritu and Arriaga's transition from the Mexican film industry to Hollywood has been well charted and regarded with ambivalence (Smith 2003, 87; Waldron 2004, 5–9; Felperin 2007, 41). On the one hand, the pair's films helped to popularize the Mexican film industry during the early 2000s, alongside those of Alfonso Cuarón and Guillermo del Toro (Waldron 2004, 5–6; Ebert 2007). On the other, Iñárritu and Arriaga's ventures into casting Hollywood actors and setting stories in America (both before and after *Babel*) intimate that Hollywood has to some extent claimed ownership of their artistic licence, so that they may be "already making films that conform to the Hollywood gaze" (Waldron 2004, 9–10). Ultimately, *Babel* and *Lantana*'s production and distribution reveal the inequities between funding and international exposure of national cinema in comparison to Hollywood's financially beneficial relationships with independent cinemas.

To a great extent, *Babel* and *Lantana* represent polar sides of the Hollywood versus national art cinema divide in terms of their production backgrounds. Despite its multiple influences, *Babel* is in its financing largely a Hollywood film, its festival presence symptomatic of the recent trend of Hollywood imperialism within the festival circuit. Yet both films show narrative concerns typically associated with art cinema. They both use network narratives to portray some of the problems of multicultural societies in an interlinked and globalizing world. They concentrate on themes of miscommunication and a lack of trust in others. These films privilege pluralism and restructure ideas about community across cultural, class and gender barriers. *Lantana* critiques the stability and sincerity of the rhetoric about Australian multiculturalism, and therein offers an interesting snapshot of contemporary suburban society. *Babel* ambitiously critiques global inequality between wealthy and developing countries, as well as problems of discrimination and miscommunication among people from different social and cultural backgrounds. Both these films place non-anglophone voices in dialogue with anglophone voices. In these ways, both films take up narrative tropes associated with art cinema and this suggests that the implied dichotomy in the term "world cinema" between popular Hollywood and other national cinemas is outdated. However, although the films use devices associated with art cinema, their narratives expound patriarchal and orientalizing values more commonly associated with mainstream cinema. This confounds the idea that Hollywood and art cinema narratives are ideologically distinct. Their practices of production, distribution and marketing illustrate that despite such a narrative blurring between industries commonly thought to be oppositional, national cinemas remain marginalized, while American companies colonize multiple channels of distribution, even in the guise of non-Hollywood films. Although these films offer themes of overcoming national differences and uneven power relationships, they reinstate such inequalities both narratively and materially.

References

Akin, Fatih, dir. 2007. *The Edge of Heaven (Auf der anderen Seite)*. Anka Film.

AllDeaf.com. 2007. "Rinko Kikuchi in *Babel*." February 25. Accessed November 20, 2010. http://www.alldeaf.com/our-world-our-culture/39291-rinko-kikuchi-babel.html

Anderson, Paul Thomas, dir. 1999. *Magnolia*. New Line Cinema.

del Mar Azcona, María. 2010. *The Multi-Protagonist Film*. Sussex: Wiley-Blackwell.

Bordwell, David. [1979] 2002. "The Art Cinema as a Mode of Film Practice." In *The European Cinema Reader*, edited by Catherine Fowler, 94-102. London: Routledge.

Bordwell, David. 2006. *The Way Hollywood Tells It: Story and Style in Modern Movies*. Berkeley and Los Angeles: University of California Press.

Bordwell, David. 2007. *Poetics of Cinema*. New York: Routledge.

Box Office Mojo. 2011. "Lantana." Accessed June 12, 2011. http://boxofficemojo.com/movies/?page=intl&id=lantana.htm&sort=todateGross&order=DESC&p=.htm

Castells, Manuel. 2000. *The Rise of the Network Society: The Information Age: Economy, Society, and Culture*. Vol. 1. 2nd ed. Malden, West Sussex: Wiley-Blackwell.

Chan, Felicia. 2011. "The International Film Festival and the Making of a National Cinema." *Screen* 52 (2): 253–260. doi: 10.1093/screen/hjr012.

Collins, Felicity, and Therese Davis. 2004. *Australian Cinema After Mabo*. Cambridge: Cambridge University Press.

Cooke, Paul, ed. 2007a. *World Cinema's 'Dialogues' with Hollywood*. Hampshire and New York: Palgrave Macmillan.

Cooke, Paul. 2007b. "Introduction." In *World Cinema's 'Dialogues' with Hollywood*, edited by Paul Cooke, 1–16. Hampshire and New York: Palgrave Macmillan.

Cooper, Merian C., and Ernest B. Schoedsack, dirs. 1933. *King Kong*. RKO.

Davies, Jude. 2005. "'Diversity. America. Leadership. Good over Evil'. Hollywood Multiculturalism and American Imperialism in *Independence Day and Three Kings*." *Patterns of Prejudice* 39 (4): 397–415. doi: 10.1080/00313220500347840.

Dennison, Stephanie, and Song Hwee Lim, eds. 2006. *Remapping World Cinema: Identity, Culture and Politics in Film*. London: Wallflower Press.

Devereaux, Mary. 1990. "Oppressive Texts, Resisting Readers and the Gendered Spectator: The New Aesthetics." *The Journal of Aesthetics and Art Criticism* 48 (4): 337–347. http://www.jstor.org/stable/431571

Duncanson, Kirsty. 2009. "The Scene of the Crime: The Uneasy Figuring of Anglo-Australian Sovereignty in the Landscape of *Lantana*." *Law Text Culture* 13 (1): 25–54. http://search.informit.com.au.virtual.anu.edu.au/fullText;dn=201001903;res=APAFT

Duncanson, Kirsty, Catriona Elder, and Murray Pratt. 2004. "Entanglement and the Modern Australian Rhythm Method: *Lantana's* Lessons in Policing Sexuality and Gender." *PORTAL* 1 (1): 1–19. http://epress.lib.uts.edu.au/journals/index.php/portal/article/view/47.

Ebert, Roger. 2002. "*Lantana*." *Chicago Sun-times*. January 18. Accessed September 12, 2008. http://rogerebert.suntimes.com/apps/pbcs.dll/article?AID=/20020118/REVIEWS/201180303/1023

Ebert, Roger. 2007. "*Babel*." *Chicago Sun-times*. September 22. Accessed July 27, 2009. http://rogerebert.suntimes.com/apps/pbcs.dll/article?AID=/20070922/REVIEWS08/70922001/1023

Emmerich, Roland, dir. 1996. *Independence Day*. Centropolis Entertainment.

Everett, Wendy. 2005. "Fractal Films and the Architecture of Complexity." *Studies in European Cinema* 2 (3): 159–171. doi: 10.1386/seci.2.3.159/1.

Faulks, Sebastian. 2011. *A Week in December*. London: Vintage.

Felperin, Leslie. 2007. "*Babel*." *Sight & Sound* 17 (2): 41–43. http://search.proquest.com.virtual.anu.edu.au/docview/237118725?accountid=8330.

Film Finance Corporation Australia. 2003. "Inquiry into Future Opportunities for Australia's Film, Animation, Special Effects and Electronic Games Industries." *Submission to The House of Representatives Standing Committee on Communications, Information Technology and the Arts*. July. Accessed June 12, 2010. http://www.aph.gov.au/house/committee/cita/film/subs/sub070.pdf

Frazer, Bryant. 2006. "*Babel* (2006)". *Deep Focus*. October 27. Accessed August 8, 2009. http://www.deep-focus.com/dfweblog/2006/10/babel_2006.html

Haggis, Paul, dir. 2004. *Crash*. Bob Yari Productions.

Halle, Randall. 2008. *German Film after Germany: Toward a Transnational Aesthetic*. Illinois: University of Illinois Press.

Haneke, Michael, dir. 2000. *Code Unknown* (Code Inconnu). Madman.

Hardy, Rod, dir. 2007. *December Boys*. AFFC.

Hassapopoulou, Marina. 2008. "*Babel*: Pushing and Reaffirming Mainstream Cinema's Boundaries." *Jump Cut: A Review of Contemporary Media* 50: 1–4. Accessed February 17, 2009. http://www.ejumpcut.org/archive/jc50.2008/Babel/index.html

Hicks, Scott, dir. 2009. *The Boys Are Back*. AFFC.

Higbee, Will, and Song Hwee Lim. 2010. "Concepts of Transnational Cinema: Towards a Critical Transnationalism in Film Studies." *Transnational Cinemas* 1 (1): 7–21. doi: 10.1386/trac.1.1.7/1.

IMDb. 2011a. "Box Office/Business for *Lantana*." Accessed June 12. http://www.imdb.com/title/tt0259393/business

IMDb. 2011b. "Release dates for *Babel*." Accessed June 13. http://www.imdb.com/title/tt0449467/releaseinfo

Iñárritu, Alejandro Gonzales, dir. 2006. *Babel*. Paramount Pictures.

Kaplan, E. Ann. [1983] 1988. *Women and Film: Both Sides of the Camera*. London and New York: Routledge, 1988.

Kerr, Paul. 2010. "*Babel*'s Network Narrative: Packaging a Globalized Art Cinema." *Transnational Cinemas* 1 (1): 37–51. doi: 10.1386/trac.1.1.37/1.

Lawrence, Ray, dir. 2001. *Lantana*. AFFC.

Lawrence, Ray, dir. 2006. *Jindabyne*. April Films.

MacFarlane, Brian. 2002. "Local/Global." *Meanjin* (March): 159–164. http://search.informit.com.au.virtual.anu.edu.au/documentSummary;dn=038446487509821;res=IELLCC

MBP. 2011. Accessed October 4. http://translate.google.com.au/translate?hl=en&sl=de&u=http://www.mbp-medien.de/&ei=CWcLT5_7NeuXiAfB-uz5BQ&sa=X&oi=translate&ct=result&resnum=2&ved=0CCkQ7gEwAQ&prev=/search%3Fq%3Dmbp%2Bfilm%2Bde%26hl%3Den%26prmd%3Dimvns

Mezgarne. 2007. "*Babel* (disappointment)." *Oasis de Mezgarne: La Gazette du Chergui*. February 6. Accessed June 11, 2009. http://translate.google.com.au/translate?hl=en&sl=fr&tl=en&u=http%3A%2F%2Fwww.mezgarne.com%2Fmaroc%2Fblog%2Fbabel-deception%2C2007%2C02

Miller, Toby. 2005. *Global Hollywood 2*. 2nd ed London: BFI Publishing.

Mitchell, David. 2004. *Cloud Atlas*. London: Hodder & Stoughton.

Naficy, Hamid. 2008. "From Accented Cinema to Multiplex Cinema." Seminar given at *FHAANZ 2008: Remapping Cinema, Remaking History*. University of Otago Film and History Association of Australia and New Zealand: November 27. http://www.otago.ac.nz/fhaanz2008/naficyabstract.PDF

Nagib, Lúcia. 2006. "Towards a Positive Definition of World Cinema." In Dennison and Lim, eds, 30–37.

Neale, Steven. 1981. "Art Cinema as Institution." *Screen* 22 (1): 11–40. doi: 10.1093/screen/22.1.11.

O'Regan, Tom. 1996. *Australian National Cinema*. London and New York: Routledge.

Orrenius, Pia M. 2001. "Illegal Immigration and Enforcement Along the U.S.–Mexico Border: An Overview." *Economic and Financial Review* First Quarter: 2-11. http://search.proquest.com.virtual.anu.edu.au/docview/196531804?accountid=8330

Pisters, Patricia 2011. "The Mosaic Film – An Affaire of Everyone: Becoming-Minoritarian in Transnational Media Culture." Accessed March 6. http://home.medewerker.uva.nl/m.g.bal/bestanden/Pisters%20Patricia%20Encuentro%20Migratory%20Politics%20READER%20OPMAAK.pdf

Quart, Alissa. 2005. "Networked: Don Roos and *Happy Endings*." August 1. Accessed August 20, 2008. http://www.alissaquart.com/articles/2005/08/networked_don_roos_and_happy_e.html

Sagal, Boris, dir. 1971. *The Omega Man*. Warner Bros. Pictures.

Scott, Allen. 2004. "Hollywood and the World: The Geography of Motion-Picture Distribution and Marketing." *Review of International Political Economy* 11 (1): 33–61. doi: 10.1080/0969229042000179758.

Shaheen, Jack G. 2003. "Reel Bad Arabs: How Hollywood Vilifies a People." *The ANNALS of the American Academy of Political and Social Science* 588 (1): 171–193. doi: 10.1177/0002716203588001011.

Silvey, Vivien. 2009. "Not Just Ensemble Films: Six Degrees, Webs, Multiplexity and the Rise of Network Narratives." *Forum: University of Edinburgh Postgraduate Journal of Culture and the Arts* (8). http://www.forumjournal.org/site/issue/08/vivien-silvey

Smith, P. J. 2003. *Amores Perros*. London: BFI.

Stam, Robert, and Louise Spence. 1983. "Colonialism, Racism and Representation." *Screen* 24 (2): 2–20. doi: 10.1093/screen/24.2.2.

Virdi, Jyotika. 2003. *The Cinematic ImagiNation: Indian Popular Films as Social History*. Delhi: Permanent Black.

Waldron, John V. 2004. "Introduction: Culture Monopolies and Mexican Cinema: A Way Out?" *Discourse* 26 (1–2): 5–25. http://search.proquest.com.virtual.anu.edu.au/docview/212435377?accountid=8330.

Willemen, Paul. 2005. "For a Comparative Film Studies." *Inter-Asia Cultural Studies* 6 (1): 98–112. doi: 10.1080/1464937042000331553.

Wood, Catherine. 2007. "Do Women Want Too Much?: *Lantana*." *Metro* 155: 147–149. http://search.informit.com.au/fullText;dn=200802250;res=APAFT.

yndprod-2. 2006. "A Serious, Thought-Provoking, Uncompromised Film ... From Hollywood?" In "Reviews and Ratings for *Babel*." *IMDb.com*, September 27. Accessed July 25, 2012. http://www.imdb.com/title/tt0449467/reviews?start=0

The global and the postcolonial in post-migratory literature

Ahmed Gamal

Ain Shams University, Cairo, Egypt

In the last 30 years of a new global era marked by fluidity and border crossing, migrant literature has proved an important literary subgenre of postcolonial writing. While the analytical category of "migrant literature" has superseded canonical categories such as "western literature" and "world literature", debate continues about its nature as a historically rooted counter-discourse to the prevailing ideologies of the global and the national. Central to the discussion of migrant literature is Elleke Boehmer's attempt to redefine such hybrid texts of the early 2000s as "post-migratory" rather than simply "migrant" in her second edition of the influential *Colonial and Postcolonial Literature: Migrant Metaphors*. This article seeks to elaborate on the main features of post-migratory literature as represented by the post-9/11 writing of Anglo-Pakistani writers, namely Mohsin Hamid's *The Reluctant Fundamentalist* (2007) and Kamila Shamsie's *Burnt Shadows* (2009). Far from being passively collusive with globalization, this new postcolonial subgenre can be defined as one mainly premised on the double agenda of either resisting neo-imperial designs or contrastively constructing transcultural contact zones grounded in reconcilable compatibilities. In this sense, post-migratory literature can offer a definitive new perspective on the borderline of the global and the postcolonial.

In June 2010 *The Daily Telegraph* published an indicative list of the best 20 British writers who represent the defining literary voices of Britain, including immigrant writers such as Mohsin Hamid and Kamila Shamsie and non-immigrant writers such as China Miéville and Steven Hall. What is remarkable is how the parochial sense of the nation is subverted by *The Daily Telegraph*'s disregard of the criterion of the country of origin in defining the new sense of national literature and hence of national identity. As Lorna Bradbury points out, "we haven't controlled the types of writing, or worried about whether writers stand in some way for different experiences of Britishness" (n. pag). With a view to Hamid's 2007 complaint about being referred to as "a Pakistani novelist" despite holding full British citizenship (qtd in Perlez n. pag), *The Daily Telegraph*'s inclusion of immigrant writers in the category of Britishness can be considered as part of a significant cultural turn that occurred in the 21st century with regard to the definition of the major distinctions between national and international, global and local. This perspective pinpoints an overwhelming drive to locate the new sense of Britishness in the contact zone where both the global and local merge harmoniously in what is known as "post-migratory literature". *The Daily Telegraph*'s list tries hard, in other words, to redefine British national literature and to thereby revise ideas of national belonging by adopting a deconstructionist approach which, unlike a multiculturalist one, presupposes the construction of culture beyond any essentialist features of the tribe in terms of gender, race or ethnicity. Consequently, *The Daily Telegraph*'s "Britain's best writers" list seems to be driven by talent and potential rather than by class or race markers.

Central to the scholarly debates on the revision of the politics of national inclusion and identity fetishism in the fields of cultural and postcolonial studies is Homi Bhabha's theorization of the "unhomely", in which the concept of postcolonial migration is fundamentally predicated upon "a process of displacement and disjunction that does not totalize" (5). With its emphases on detotalization and reconfiguring the border between home and the world, Bhabha's unhomeliness hypothesis, thus defined, accords very obviously with the post-migratory condition which is similarly "inherent in that rite of extra-territorial and cross-cultural initiation" (9). One question that needs to be asked, however, is whether the unhomely intellectual attempts to syncretize or disrupt both the local and the global. Abdul R. Jan Mohamed's typology of the "specular border intellectual" and the "syncretic border intellectual" is relevant here. According to his definition, a post-migratory perspective can correspond to the double vision of syncretic border intellectuals who find themselves caught between two cultures and yet are "able to combine elements of the two cultures in order to articulate new syncretic forms and experiences" (97). In other words, those post-migratory writers or border intellectuals negotiate their relationship to the nation of the periphery (i.e. that of their birth), the history that precedes and threatens to determine them, and the centred, western canon into which they strive to write themselves.

What is post-migratory literature?

There is a growing tendency in the US academy to classify contemporary fiction according to the categories of exile and migrancy as either "home" (of a national affiliation) or "diasporic" (of a transnational affiliation) (Lau 237). In these geographically-oriented analyses, the "diasporic" and "immigrant" writers enjoy an advantage in terms of their global and multicultural profile. Such classifications depend on either the geographic stratification of several sites of production, circulation and translation, or the linguistic categorization of different strata of bilingualism and multilingualism. The institutionalization of new writings in English is thus still circumscribed by the traditional idiosyncrasies of the national, the geographic and the lingual. The deterritorialized writing of anglophone South Asian writers such as Aamer Hussein, Kamila Shamsie and Mohsin Hamid, anglophone Caribbean writers such as Wilson Harris, George Lamming and V.S. Naipaul, and anglophone African writers such as Ahdaf Soueif, Ben Okri and Ayi Kwei Armah would be hence reterritorialized and conceptualized as writing which represents new dimensions in the British and South Asian, Caribbean and African cultural scenes rather than the global or the postcolonial context in general.

By contrast, Carine Mardorossian's concept of "migrant literature" is predicated upon a typically counter-traditional ontological paradigm of migrancy which can supposedly encompass all new writings, from within both the western metropolis and the postcolonial peripheries and by both immigrant and non-immigrant writers, that focus on the quest for alternative modernities in the context of the new world order and globalization. In contrast to a literature with a nationalist ethos, "migrant literature" is grounded in a cosmopolitan perspective that seeks to disrupt the binary logic of modernization and tradition, of origin community and host country. According to such a paradigm, "the reader's conventional idea of home needs to be revised to include a more complex conceptualization, one which foregrounds ambivalence, fragmentation, and plurality as a new way of thinking about space and identity" (Mardorossian 22). In other words, the idea of "home" is redefined as a social reality that is structured by discourses. The blanket category of "migrant literature" can thus distinguish new writings as coherent and essentially concerned with liminal and borderline existences, both home and abroad. Its signification is no longer confined to the expatriate experience or to the social context of the writers or their publication and translation milieus, but is rather associated with an aesthetic programme that might be equally shared by all types of writers. In the words of Rebecca Walkowitz, "Not every book that travels is produced by a writer who travels" (532). By the same token, the category of "post-migratory literature" is hypothetically characterized by a double movement across the global and the postcolonial. Far from being static or essentialist, this movement is a dynamic one, grounded in constant renegotiations and reconstructions.

Thus, post-migratory literature can be described as that type of postcolonial literature that fundamentally problematizes the condition of migrancy by deconstructing the binarism of home and the world and linking the global to the postcolonial. The "post" therefore represents an oppositional rhetoric of emerging voices which are profoundly contestatory of the hierarchy of binaristic essentialism. In response to the post-9/11 nostalgic discourse of the dominant which excludes difference and heterogeneity as terror or impurity, post-migratory literature offers textual models of hybridization and dialogic exchange as well as resistance and liberation. In Elleke Boehmer's words:

> Such "post-migratory" "post-postcolonial" writing, as "post-migratory" black British writer Caryl Phillips terms it, explores not only leave taking and departure, watchwords of the migrant condition, but also the regeneration of communities and selves out of heterogeneous experiences in the new country. (250)

The reconstruction of new contact zones within the historical settings and transnational context of post-migratory narratives is hence styled as fundamentally cosmopolitan. However, the common "post" in post-migratory and postcolonial literature connotes a manifest oppositional stance that might be unavailable in unconditional cosmopolitanism. In addition, post-migratory writings deftly portray characters that retain their sense of tradition, cultural heritage and, accordingly, their postcolonial backgrounds in their newly adopted homelands. The bilateralism of the global and the postcolonial is thus proposed as the most significant and typical feature of the post-migratory literature which thematizes migrancy in both its form and content. This is what basically distinguishes post-migratory literature from the migrant literature of the 1980s and the 1990s which "remains collusive with and an expression of the neo-colonial world" (Boehmer 238).

Both Mohsin Hamid and Kamila Shamsie are particularly suited to serve as representative instances of post-migratory literature with respect to their real and fictional worlds. Hamid, born in Pakistan and educated in both the US and Pakistan, graduated from Princeton University in 1993 and now divides his time between Pakistan and abroad, living between Lahore, New York, London, and Mediterranean countries such as Italy and Greece. Similarly, Pakistan-born Shamsie has an MFA from the MFA Program for Poets and Writers at the University of Massachusetts, works as a reviewer and columnist for *The Guardian* and lives in London and Karachi. Hamid and Shamsie thus belong to that group of Pakistani writers in English who "neither have hyphenated identities nor can be considered Pakistani exiles, but write in liminal positions between West and East" (Chambers 124).

The very concepts of unhomely identities and a post-migratory imaginary – as grounds of cultural comparativism – are particularly relevant to the experience of the Pakistani diaspora which is metonymic of the new South Asian diaspora in general. The traumatic rite of passage to the respective postcolonial nation states of India, Pakistan and later Bangladesh that emerged in the aftermath of the Partition consolidated the emergent dialectical narratives of homecoming and leave-taking as founding metaphors in the modern alternative histories of the nation. The most significant effect of this migrant process was the proliferation of the liminal space of an expansive South Asian identity and its consequent affiliation with a transnational sensibility resulting from one of the world's largest diasporas. In a recent issue of *South Asian Review*, South Asian diaspora studies are spoken of as mainly interested in how

> the South Asian idea functions not only as a cognitive instrument designed to suture a myriad of differences that are commonplace in the region but also a strategic speech act that calls forth a more encompassing transnational regional identity as a better alternative to the current tyranny of the nation-centric (to use the title of a popular Amitav Gosh novel) "shadow lines". (Giri and Kumar 14)

Both *The Reluctant Fundamentalist* and *Burnt Shadows* hence abound with examples of a transnational sensibility that is comprised of South Asian and similar third world post-migratory imaginaries.

Another notable aspect is that a post-migratory imaginary is hypothesized to be radically different from the classical constructs of diaspora that are associated with loss, trauma, exile and suffering (Cohen 3). As demonstrated in the narrative strands of migrancy in the two novels (by Changez in *The Reluctant Fundamentalist* and both Hiroko and Raza in *Burnt Shadows*), a post-migratory stance is basically grounded in a sense of survival and reshaping. In other words, the migrant strands do not conceptualize the homeland as a place of origin and site of departure that constitutes a certain people as diasporic. On the contrary, the homeland, as a utopian destination for South Asian diasporics, is constructed as only one element of the diasporic imaginary and thus must be understood as a temporal and cultural process rather than a place.

What is particularly new about post-migratory literature produced by writers like Soueif, Desai, Ghosh, Hamid and Shamsie is its proclivity toward a dynamic and undeviating reconstruction of the global and the postcolonial. The aesthetic structure of post-migratory novels thus tends to appropriate a rather more circular form of Bildungsroman, replete with recurrent continuities and discontinuities across global and postcolonial geographies and histories. Conversely, migrant literature of the 1980s and 1990s endorsed a more linear form of Bildungsroman, where the protagonist starts in a romanticized version of the host country and ends in disillusionment in his original home country. In such works, the gap between the global and the postcolonial was hence constructed as hardly bridgeable and translatable. V.S. Naipaul's *The Enigma of Arrival* (1987) is a good example of such diasporic writing; it offers an intellectual autobiography of a migrant writer who is portrayed as infatuated with the English countryside, represented by Jack's garden at the beginning of the novel, and then as cognizant of the sanctities of national belonging and historical weightiness only at the very end, when they are inevitably lost and remade:

Every generation now was to take us further away from those sanctities. But we remade the world for ourselves; every generation does that, as we found when we came together for the death of this sister and felt the need to honor and remember. (354)

Unlike migrant literature of the 1980s and 1990s, Hamid's *The Reluctant Fundamentalist* and Shamsie's *Burnt Shadows* are predicated on the uniform dialectical intersection of the global and the postcolonial and the construction of a post-migratory identity. The progressive motion of Hamid's Pakistani migrant protagonist, Changez, ranges across the widely separate geographical and cultural spaces of New York, New Jersey, Lahore, Rhodes, Manila and Valparaiso. Yet however immense and global the space navigated, the navigator's basic focalization is deeply rooted in his homeland. Through his reflective monologue addressed to a muffled US interlocutor, Changez draws a striking comparison between Manhattan and the older districts of Lahore at the very beginning of the novel: both are democratically urban spaces favouring pedestrians and abounding in Pak-Punjab delis and resounding to Urdu, spoken by cab drivers in New York and Pakistanis in Lahore. Therefore, moving to New York, for Changez, seems like coming home. The modern global grandeur of the US is, moreover, counterbalanced by the narrator's description of Pakistan's historical grandeur. Lahore is accordingly focalized as a symbolic icon of the historical weightiness of Pakistan: "I said I was from Lahore, ancient capital of the Punjab, home to nearly as many people as New York, layered like a sedimentary plain with the accreted history of invaders from the Aryans to the Mongols to the British" (7). In Chile, however, Valparaiso helps Changez achieve a kind of irrevocable disillusionment some time before the novel's open ending. In addition to his emergent sense of alienation after 9/11, Changez discovers that Valparaiso had been continuously in decline due to the peripheral location it came to have due to American intervention and the construction of the Panama Canal. It rather reminds him of Lahore's former aspirations to grandeur. Furthermore, he intellectually identifies himself with Valparaiso despite its geographic remoteness from his homeland; for him, both cities have the same spirit. Before that epiphany, he lacked "a stable *core*", his "identity was so fragile" (148, original emphasis), and he was consequently lost in cultural expatriation. In Valparaiso Changez manages to formulate a post-migratory identity and a transnational sensibility that comprise all non-western countries, namely Vietnam, Korea, Taiwan, Iraq, the Middle East and Afghanistan, which were constantly subjected to US coercion and interference.

On a more allegorical level, Changez negotiates his relationship to the peripheral nation of his birth and the centred, western tradition into which he strives to write himself through his love for a young American woman. Put differently, post-migratory narratives transform their border-crossing among third world cultures and simultaneously between western and non-western cultures into a positive mission that generates cultural dialogue. Changez and Erica's encounter frames the whole narrative in a larger sense, because Hamid defers the event of their sexual union and, after a few promising dates, manages to avoid it for much of the novel. Both characters meet on Rhodes, an island guarded by a wall which metaphorically stands for the cultural barriers between two different world views. Erica's final descent into depression and suicide and Changez's decision to leave the US symbolize such barriers. Despite this dilemma, part of Changez retains a cordial association with Erica, whose name echoes the final two syllables of "America": "I had returned to Pakistan, but my inhabitation of your country had not entirely ceased. I remained emotionally entwined with Erica" (172). The version of post-migratory identity that he finally acquires is based on the interracial relationship with Erica; explaining his newly constructed identity, he concedes:

> Such journeys have convinced me that it is not always possible to restore one's boundaries after they have been blurred and made permeable by a relationship: try as we might, we cannot reconstitute ourselves as the autonomous beings we previously imagined ourselves to be. (173–74)

The novel's open ending thus depicts Changez's plural identity as an incessant process that accords with the needs of both the global and the local.

Shamsie's *Burnt Shadows* is likewise engaged with a dialectical relation between the global and the postcolonial. Unlike the minimalist structure of *The Reluctant Fundamentalist*, *Burnt Shadows* is an epic family saga and a useful reminder of history as seen from non-western eyes. The main character is the Japanese woman, Hiroko Tanaka, who strives to keep connected to her Oriental origins and simultaneously to construct her identity according to new cultural locations. In *Burnt Shadows* all narratives of migrant characters are inscribed as minor stories within the frame story of loss and regeneration. The first narrative relates how Konrad Weiss leaves Nazi Germany to migrate to cosmopolitan Nagasaki at the end of the Second World War and finally loses his life. The second narrative focuses on how Sajjad Ali Ashraf, the Burtons' Muslim employee, loses his Dilli, or old Delhi, and his family in India after the Partition, and reshapes a new home and a new family in Pakistan. The third narrative involves Sajjad's son Raza, who is a true threshold figure, symbolically representing the condition of migrancy from east to west and literally the translational world of migrants in which he works as a translator attempting to downplay his manifest difference among Pakistanis and Afghans respectively. Similarly, Harry, the son of James and Ilse, is another liminal figure disabled by his foreignness in England, the US, Afghanistan and Pakistan. The climax of both *The Reluctant Fundamentalist* and *Burnt Shadows* centres on a collision of the main character's world, which adheres to Muslim values, and the US post-9/11 world, which profiles Muslims as potential terrorists.

A comparison can be drawn between Changez's story in *The Reluctant Fundamentalist* and Hiroko's in *Burnt Shadows* as regards the potential constructive agency of migrant experience. Despite Hiroko's diasporic dispersion to many foreign regions, ranging from Delhi to Istanbul, Karachi to New York, she retains a robust vision of her original homeland Nagasaki and its cosmopolitan history before the bombing on 9 August 1945, and later of her adopted homeland Karachi after her final immigration to the US. Her departure from both is associated with an inheritance of the incongruities of losing one's home and the desire to construct a new one. Although Hiroko is exiled from Japan because of the cultural coercion profiling her in the aftermath of the bomb as *hibakusha* or an explosion-affected person, she persists in reassembling her memories of Harry Truman, Nagasaki, the 75,000 dead Japanese and the bird-shaped burns on her back as a constructive strategy to transform the image of pain into forms of resistance to external oppression represented by US post-9/11 Islamophobia. At the core of Hiroko's account of the traumatic history of Japan is her own reflexive construction of a trauma-based selfhood that can be constantly born anew. Regardless of the sense of loss and nostalgia encountered by migrant characters in Shamsie's *Burnt Shadows*, each of them succeeds in critically constructing a present that rewrites a trauma-based past. Moving from Tokyo to Delhi, Hiroko feels like a figure out of myth, a "character who loses everything and is born anew in blood" (49). This ambivalence within the migrant experience is partly reflected in the doubleness of imagined communities whose cultural representation, according to Homi Bhabha, "moves between cultural formations and social processes without a centred causal logic" (141). In India, Hiroko moves to Delhi, looking for a new beginning in solidarity with Konrad Weiss's family, namely his sister Ilse or Elizabeth and her English husband James Burton. In spite of her alienation from the English and her drastic sense of the loss of her own Japanese culture, she admits that the Burtons furnished her migrant experience with something of significance: "The belief that there are worthwhile things still to be found. All I've been doing all this while is thinking of losses. So much lost" (49).

In Karachi, Hiroko's ability to meld into Sajjad's new world is similarly premised upon negotiation and reciprocity rather than rejection and hierarchy. With respect to domestic life, she makes concessions to her son's nationalist critique of her short dresses by packing them away and wearing shalwar kameezes, the Pakistani traditional loose pyjama-like trousers worn by both women and men in South Asia. In addition, her marital relationship to Sajjad is portrayed as comprising a series of continuous negotiations between her proclivity for rebelliousness and his conservative nature, between her determination to work and teach, for example, and his to provide for his wife and family. Regardless of her ambivalent stance towards home, Hiroko faces the growing realization that her life story is grounded in survival and momentum more than in desperation and loss. It is in the middle of the novel that she becomes fully aware of her agency, which seems straightforwardly reassuring and liberating: "So the story of Hiroko Ashraf's youth was not the story of the bomb, but of the voyage after it" (223).

In Hamid and Shamsie's novels, the space of post-9/11 New York is particularly inscribed as a site of potential transnational sensibility and exchange. After Hiroko's eventual departure to New York, she associates the 9/11 victims and missing people with those in Nagasaki after the bomb blast. Her feeling of solidarity is inscribed as "quite unfamiliar, utterly overwhelming" (289), to the extent that she retreats from giving blood only when strictly told that she is from a malarial country. Like *The Reluctant Fundamentalist*, *Burnt Shadows* is a circular narrative that disavows closure and hence spotlights the potential significance of border crossing and migrancy. On the whole, Hiroko's story interrogates the critical construction of the discrepancy between "migrancy" and "diaspora" in postcolonial literature. Whereas the term "diaspora" is discussed by Rosemary George, for instance, in terms of the identity space provided by history, origin, solidarity and community (182), the term "migrancy" is denied such surfeit of rootedness and solidarity. In spite of this disparaging categorization, Changez and Hiroko's struggle to maintain solidarity with their own tribe and that of the other, namely Americans, amply demonstrates how migrancy can function as an effective strategy of transformation and mobilization in contact zones constructed within post-migratory literature.

Imperial designs and postcolonial resistance

In keeping with Enrique Dussel's philosophical discourse on modernity, which differentiates between the "Eurocentric" and the "planetary" paradigms, this article proposes a distinction between the "imperial" and the "cosmopolitan" paradigms with regard to the global project as an historical extension of capitalist modernity. The Eurocentric paradigm, according to Dussel, is meant to signal a totalitarian, authoritative structure that legitimizes the universality and supremacy of European culture, "not only in Europe and in the United States, but in the entire intellectual realm of the world periphery" (3–4). The imperial paradigm has much in common with the Eurocentric one in terms of their centre-periphery polarization and totalitarian cultural unification and penetration. Revathi Krishnaswamy writes: "The merger of the American nation-state with the state of the globe is among the most significant aspects of globalization 9/11 or 'imperiality,' as I propose to call it" (12). This imperial propensity for unification and the subsequent binary thinking of "the west and the rest" and "us and them" have been significantly resuscitated after 9/11. In conformity with such binaristic grammar, Thomas Friedman, for instance, identifies polarization as foundational in the new international order after 9/11. In spite of his demonstration that a globalizing perspective is epistemologically necessary for a world without frontiers or walls through what he calls a *"flat-world platform"* (*The World* 10; original emphasis), he still conceptualizes the new world system as polarized and divided between "the World of Order and the World of Disorder" instead of being divided between east and west or north and south (Friedman, "Peking Duct Tape" par. 4). In other words, such a flat world could only be conceptualized, not experienced, because the blurring of boundaries between the west and its others might induce implacable threats and cheerless convergences. In this case, the geopolitical imaginary nourished by a post-migratory position is used as a counter to imperial designs or globalization from above.

Both *The Reluctant Fundamentalist* and *Burnt Shadows*, therefore, narrativize through social, economic and power relations the construction of imperial designs which are driven by the will to control and homogenize, and the counter-construction of postcolonial resistance and delinking. In both *The Reluctant Fundamentalist* and *Burnt Shadows* Changez and Hiroko's relation to the American way of life is problematized as simultaneously grounded in integration as well as separation. Lured by the American dream of global education, Changez conceives of Princeton as a hyper-real world that abounds with titan professors and philosopher-king students. In a mediatized world, he imagines himself as a film star rather than as an academic student. The triumph of Princeton, according to Changez, is not merely educational, but also economic and political. The customized evaluation and testing systems exemplify the political power of US pragmatism and efficiency, whereas the global financial sourcing of international students like Changez represents the strong financial system of US education. Changez critically remarks that "America had universities with individual endowments greater than our national budget for education" (34).

For Changez, Underwood Samson & Company similarly represents how difficult it is to think of the economy as either benign or as emptied of imperial power relations. With its high-tech office buildings, Changez realizes that Underwood Samson represents the power of "the most technologically advanced civilization our species had ever known" (34). Social alienation is consequently explored as a corollary of global capitalism. Changez feels alienated from the hostile employees in New Jersey, the Filipino driver in Manila and finally Juan-Bautista in Valparaiso for his own active part in global economy and corporate dreams.

The climax of the novel centres on the protagonist's epiphany with regard to the ravages of capitalist exploitation and the commodification of nature rather than on the racist discrimination against Muslims ignited in the wake of 9/11. The novel does not trace "the experience of alienation and estrangement to a particular historical trigger", as conceded by Peter Morey (139). Such a view monumentalizes 9/11 as an "absolute" historical juncture, far removed from the historicity of late capitalism, and reduces the core–periphery polarity characteristic of global capitalism to a mere clash of civilizations. Early in the novel, Changez is given some cautionary advice by his African-American colleague Wainwright about the dark side of the American dream of globalization. Although

Changez strives to act as an American in Manila, he gradually realizes that his American colleagues are just foreign to him and that he might feel closer to the Filipino driver, because both of them share a sort of "Third World sensibility" (67).

In accordance with current critical globalization studies, by invoking the strategy of "delinking", *The Reluctant Fundamentalist* conceptualizes the necessary break from the world capitalist system and the refusal to submit to the imperatives of globalization as an effective postcolonial strategy of resistance that also constitutes a fundamental consolidation of cultural identity (Amin, "Delinking" 435). Although Samir Amin discusses delinking in terms of power and economic relations only, both Hamid and Shamsie try to associate the political and the economic with the cultural by contesting western provisional constructions of third world dependency and atrophy. In Chile, Changez discovers the brutality of the global differential distribution of power due to the help of Juan-Bautista, the chief of the publishing company to be valued by Underwood Samson. Juan-Bautista directs Changez to Pablo Neruda's house, an allusion that falls in line with the anti-imperial stance adopted by Neruda and eventually by Changez himself. Juan-Bautista succeeds in demystifying the actual relation between the empire and its adopted soldiers or janissaries. Changez becomes cognizant of the close affinity between the janissaries who were Christian boys recruited in the Ottoman army to fight against their own civilization on the one hand, and the third world business officers employed by multinational companies to disrupt the lives of the deprived and the dehumanized on the other:

> I had thrown in my lot with the men of Underwood Samson, with the officers of the empire, when all along I was predisposed to feel compassion for those, like Juan-Bautista, whose lives the empire thought nothing of overturning for its own gain. (152)

Changez's final resolution to quit Underwood Samson and return to Pakistan seems to concretize the quintessence of postcolonial politics, namely the preoccupation with "those whose knowledges and histories are not allowed to count" (Young 14).

Burnt Shadows similarly portrays Hiroko's relation to the US imperial dream as ambivalent. Imperial war is delineated in the first part of the novel as an inherently violent and destructive phenomenon; it results in the incineration of both Nagasaki's cosmopolitanism and Hiroko's personal dreams of becoming a "modern girl" living in a big city, wearing dresses and enjoying the lifestyle of jazz clubs. Regardless of Hiroko's admiration of American informality, the bomb haunts her whole life through the burnt shadows on her back. Like Changez, Hiroko nevertheless identifies with an American lifestyle in Tokyo: she works as a translator for Americans; she befriends an American nurse and borrows her clothes; she goes out to nightclubs and has her hair cut short like American young women. In spite of such growing affinity with Americans, she decides to delink herself from that world when she is told that the bombing had to be done to save American lives.

Like Changez, Raza is employed by the empire as a janissary against his people after 9/11. As Sajjad is employed by James Burton, Raza is appointed by an US security corporation as Harry Burton's assistant. Harry is himself aware of the similarity of global imperial designs regardless of their different geopolitical contexts. His colleague Steve jokes that the only difference between Vietnam and Afghanistan is that they have GI in the first and "jee-had" in the second. Just as Changez is selected by Jim to be employed to the advantage of the covert economic operations of the empire in *The Reluctant Fundamentalist*, Raza is elected by Harry for his superb translation and reshaping skills which so convince the Afghans that he is one of them that they take him to a mujahideen camp in *Burnt Shadows*. In addition, Raza's final reluctance to work for Arkwright and Glenn has a conspicuous similarity to Changez's unwillingness to work for Underwood Samson. Regardless of the fact that Raza lives in Miami for a decade and is a green-card holder, he conceives of the US as unluckily jangling with fear of terrorism and re-mythologizing 9/11 as "the absolute event, the 'mother' of all events" in Jean Baudrillard's words (4). Delinking himself from global war at the climax of the novel, Raza decides to help Abdullah after running from the FBI in the wake of 9/11 by first asking Kim to smuggle him across the border and then by consciously acting out Abdullah's personality.

By critiquing the dynamics of imperial supremacy, post-migratory literature builds on and extends a long-standing liberating strategy in what is theoretically known as postcolonial cosmopolitanism. According to Revathi Krishnaswamy, "postcolonial cosmopolitanism appears to work against all forms of totalization and homogenization, be it modernization, Westernization or Americanization, capitalism, or nationalism" (3). The crucial point here is that the fundamentalist reassertion of an authoritative national identity is the product of neocolonial globalization. The discourses of both flag nationalism and religious fundamentalism thus have the same totalizing constructs of globalization and the same "all-encompassing aims of explaining history and nature" (Amin and Luckin 222). Thus, fundamentalism is critiqued in both *The Reluctant Fundamentalist* and *Burnt Shadows* in terms of its economic, political and religious variances.

The title of Hamid's *The Reluctant Fundamentalist* reflects the ideological schizophrenia of first world globalism as well as third world nationalism that keeps cosmopolitan citizens like Changez in emotional conflict with themselves regarding their self-worth. An ironic tone is hence adopted with regard to both Underwood Samson's pursuit of economic "fundamentals" and Changez's dream to become "the dictator of an Islamic republic with nuclear capability" (29). Changez's schadenfreude in response to the painful scene of the fall of the towers is equally compared to a US citizen taking pleasure in the laying waste of the structures of America's enemies. Chauvinism on both fronts is satirized for being grounded in bloodthirsty and apathetic rhetoric. Historically speaking, both British colonialism and religious nationalism are comparatively constructed as detrimental in *Burnt Shadows*; Indian Muslims who left India for Pakistan are delineated as leaving their home, while the British who had not gone native like the Turks, Arabs, Mongols and Persians are portrayed as resistant to entering an India outside the Raj. Moreover, the violence caused by the Partition in 1947 is described as creating an atmosphere of mutual hostility and suspicion between India and Pakistan that plagues their relationship to this day. Sajjad himself feels it is wrong to be touched by atrocities committed on Muslims more than by those committed by Muslims. The final message drawn from the two novels is: "to combat a single-minded and ruthless fanaticism by becoming equally fanatical and ruthless […] will not further the cause of justice or bring about a meaningful democracy. It can only prolong the cycle of violence" (Ali 3).

Cosmopolitan designs and postcolonial contact zones

According to Timothy Brennan, cosmopolitanism "designates an enthusiasm for customary differences, but as ethical or aesthetic material for a unified polychromatic culture – a new singularity born of a blending and merging of multiple local constituents" (41). Both cosmopolitanism and post-migrancy celebrate difference and hybridity. Post-migratory texts conceivably benefit from an engagement with the cosmopolitan acknowledgement of difference, deconstruction of sameness and the preoccupation with liminal spaces and contact zones. However, such cosmopolitan convergence of cultures is constructed in both *The Reluctant Fundamentalist* and *Burnt Shadows* as either conscious or unconscious. Forced or unconscious cosmopolitanism is hence associated with the flows of migration, production and consumption into urban centres, whereas conscious cosmopolitanism is grounded in active reordering and restructuring of the world and identity. At the core of the construction of cosmopolitan identity in the two novels is an understanding of subjectivity as bound up with the images of narration and translation. This metafictional alignment of subjectivity and story is fundamental to contemporary theories of identity. As Jerome Bruner argues, "we organize our experience and our memory of human happenings mainly in the form of narrative" (4). Thus, both Hamid and Shamsie rewrite their recommended version of cosmopolitanism by constructing personal and social narratives of reciprocity and conviviality, negotiation and translation.

Changez and Erica's love story in *The Reluctant Fundamentalist* represents the unexplored potential of another American narrative, one that might acknowledge the latent potency of the third world's postcoloniality and the Arab Spring in the Middle East. For Anna Hartnell, the narrative "apparently yearns for an 'other' America; one that, like Erica, occupies an 'otherworldly' space, but a space that might recognize Changez as a fellow bearer of a conflicted postcolonial legacy" (345). Erica is first introduced in the novel as a narrative persona removed from external reality and more associated with American cinematography in terms of contemporary female iconic figures such as Gwyneth Paltrow and Britney Spears. Changez's admiration for her nudity in Greece is further linked to a literary allusion to Italo Calvino's *Mr Palomar*: a text that incorporates the same symbolic semblance between the natural and the cultural, the real and the fictional.

In the aftermath of 9/11, migrant characters are excluded from the narrative of US conviviality and openness. The double tropes of movie watching and fiction writing are spotlighted with regard to the isolation of foreigners compared to living in a black and white film about the Second World War on the one hand, and to the missing episode of migrants in the American script on the other. Changez accordingly feels spurned by both Erica and the USA at large. Comparatively, Erica actually fails to draw Changez to her narrative of the New York elitist world and to reconcile him to her father's "typically *American* undercurrent of condescension" (55). Though Changez identifies himself with Erica and Chris's love story and their commingling of identities, he fails to let her forget about Chris or her past. However, Changez remains dedicated to Erica's memory after returning to Pakistan, a fact that reflects the importance post-migratory writing places on a cosmopolitan sense deeply rooted in the home country.

Similarly, in *Burnt Shadows*, the cosmopolitan reality of the Weiss-Burtons and the Tanaka-Ashrafs is also constructed in relation to narrativity. Their story is consequently compared to the Islamic narrative of the spider, told by different members of the two families, in which the spider is represented as a metafictional reference associating the Prophet of Islam's immigration and survival with the two families' mutual devotion and transformation via migrancy.

Changez and Erica's story is also initiated by a penchant for cultural translation; Changez's elaborate and impassioned accounts of the rich Pakistani geography and his attempt to write Erica's name in Urdu underscore Hamid's keen pursuit of cross-cultural exchange and cultural translation. "As in ethnographic studies", Ruvani Ranasinha writes, "postcolonial fiction and criticism foreground the spatial mapping of cultural translation, especially in relation to migrancy, and tend to erase the complex question of the temporal" (5). Convivial reciprocity between the Weiss-Burtons and the Tanaka-Ashrafs is comparatively demonstrated in relation to cultural translation in *Burnt Shadows*. Tellingly,

Hiroko becomes attached to Konrad when she starts translating Japanese letters for his planned book about the cosmopolitan world of Nagasaki, replete with descriptions of Euro-Japanese buildings, international clubs and intermarriages. In Delhi, Hiroko becomes deeply involved with Sajjad through the Urdu teaching sessions provided by the Burtons' cosmopolitan generosity.

Aesthetically, the post-migratory stance of Hamid and Shamsie is reflected in the alternating currents of Changez's single narrative voice in *The Reluctant Fundamentalist* and the multiple, shifting focalization in *Burnt Shadows*. The two novels, furthermore, blend western realism with eastern mysticism and lyricism. The two texts also adopt the more postcolonial and critical stance of cosmopolitanism which spotlights its attachment to local history and consciousness. According to Bruce Robbins, cosmopolitanism should be defined as habits of thought that have been shaped by particular collectivities, "that are socially and geographically situated" (2). Their historical allusions are accordingly foregrounded to underline the particular sense of the past grandeur of Pakistan and the nostalgia for its recuperation. In *The Reluctant Fundamentalist* Changez longs for a time when the people of the Indus River basin had an orderly development of settlements and communities, whereas Sajjad's allusion in *Burnt Shadows* to Rani of Jhansi, one of the leading figures of the Indian Rebellion of 1857 and a symbol of resistance to British rule in India, and Razia of the Mamluk Dynasty, the first female ruler in the Muslim world in medieval India, draws attention to the past ideal of female empowerment in the east.

This article has explained the central importance of the categories of the global and the postcolonial in post-migratory literature, which offer a wide range of literary responses to the concern with constructing imagined links between tradition and modernity, the past and the present, and the postcolonial and the global. It suggests that these links comprise the double strategies of linking and delinking; that the global and the postcolonial, as analytical categories, cannot be isolated from the political and historical experience of different characters. While the stress on categories such as diaspora and migrancy remains valid and necessary in the study of new postcolonial writing, such a focus, when excluding those of the global and the postcolonial, falls short of accounting for the complexity and political uniqueness of post-migratory writing.

Works cited

Ali, Tariq. *The Clash of Fundamentalisms: Crusades, Jihads and Modernity.* London: Verso, 2002.

Amin, Samir. "A Note on the Concept of Delinking." *Review* 10.3 (1987): 435–44.

Amin, Samir, and David Luckin. "The Challenge of Globalization." *Review of International Political Economy* 3.2 (1996): 216–59.

Baudrillard, Jean. *The Spirit of Terrorism and Requiem for the Twin Towers.* Trans. Chris Turner. London: Verso, 2002.

Bhabha, Homi K. *The Location of Culture.* London: Routledge, 1994.

Boehmer, Elleke. *Colonial and Postcolonial Literature: Migrant Metaphors*. 2nd ed. Oxford: Oxford UP, 2005.

Bradbury, Lorna. "Are these Britain's Best 20 Novelists under 40?" *Daily Telegraph* 18 Jun 2010. Telegraph.co.uk. Web. 18 Jun 2010.

Brennan, Timothy. "Cosmopolitanism and Internationalism." *Debating Cosmopolitics,* Ed. Archibugi Daniele. London: Verso, 2000.

Bruner, Jerome. "The Narrative Construction of Reality." *Critical Inquiry* 18.1 (1991): 1–21.

Chambers, Claire. "A Comparative Approach to Pakistani Fiction in English." *Journal of Postcolonial Writing* 47.2 (2011): 122–34.

Cohen, Robin. *Global Diasporas: An Introduction.* Seattle: U of Washington P, 1997.

Dussel, Enrique. "Beyond Eurocentrism: The World-System and the Limits of Modernity". *The Cultures of Globalization.* Ed. Fredric Jameson and Massao Miyoshi. Durham, NC: Duke UP, 1998. 3–31.

Friedman, Thomas L. "Peking Duct Tape". *New York Times.*16 February 2003. nytimes.com. Web. 16 February 2003.

———. *The World Is Flat: A Brief History of the Twenty-First Century.* New York: Picador, 2005.

George, Rosemary Marangoly. " 'At a Slight Angle to Reality': Reading Indian Diaspora Literature". *MELUS* 21.3 (1996): 179–93.

Giri, B.P., and Priya Kumar. "On South Asian Diasporas". *South Asian Review* 32.3 (2011): 11–26.

Hamid, Mohsin. *The Reluctant Fundamentalist.* Boston: Mariner Books, 2007.

Hartnell, Anna. "Moving through America: Race, Place and Resistance in Mohsin Hamid's *The Reluctant Fundamentalist.*" *Journal of Postcolonial Writing* 46.3–4 (2010): 336–48.

Jan Mohamed, Abdul R. "Worldliness-without-World, Homelessness-as-Home: Toward a Definition of the Specular Border Intellectual". *Edward Said: A Critical Reader.* Ed. Michael Sprinker. Oxford: Blackwell, 1992. 96–120.

Krishnaswamy, Revathi. "Postcolonial and Globalization Studies: Connections, Conflicts, Complicities". *The Postcolonial and the Global.* Ed. Revathi Krishnaswamy and John C. Hawley. Minneapolis: U of Minnesota P, 2008. 2–21.

Lau, Lisa. "Making the Difference. The Differing Presentations and Representations of South Asia in the Contemporary Fiction of Home and Diasporic South Asian Women Writers." *Modern Asian Studies* 39.1 (2005): 237–56.

Mardorossian, Carine. "From Literature of Exile to Migrant Literature." *Modern Language Studies* 32.2 (2002): 15–33.

Morey, Peter. "'The Rules of the Game have Changed': Mohsin Hamid's The Reluctant Fundamentalist and Post-9/11 Fiction." *Journal of Postcolonial Writing* 47.2 (2011): 135–46.

Naipaul, V.S. *The Enigma of Arrival.* New York: Vintage Books, 1987.

Perlez, Jane. "Mohsin Hamid: A Muslim Novelist's Eye on US and Europe". *New York Times* 12 October 2007, nytimes.com. Web. 12 October 2007.

Ranasinha, Ruvani. *South Asian Writers in Twentieth-Century Britain: Culture in Translation.* Oxford: Clarendon P, 2007.

Robbins, Bruce. "Actually Existing Cosmopolitanism". *Cosmopolitics: Thinking and Feeling beyond the Nation.* Ed. Pheng Cheah and Bruce Robbins. Minneapolis: U of Minnesota P, 1998. 1–19.

Shamsie, Kamila. *Burnt Shadows.* London: Bloomsbury, 2009.

Walkowitz, Rebecca L. "The Location of Literature: The Transnational Book and the Migrant Writer." *Contemporary Literature* 47.4 (2006): 527–45.

Young, Robert J.C. "What is the Postcolonial?" *Ariel: A Review of International English Literature* 40.1 (2009): 13–25.

The cartography of the local in Arun Kolatkar's poetry

Anjali Nerlekar

Rutgers University, USA

The post-independence bilingual Indian poet Arun Kolatkar (1932–2004) uses cartographic images and narratives of travel to interrogate the newly independent Indian nation. Focusing on the city of Bombay/Mumbai, the poet persistently maps the city and its environs in unconventional ways: on foot, through eating habits and clothing. Such walking documentation of city spaces provides a resistant alternative to privileged viewpoints of spaces and people, as de Certeau points out, and Kolatkar's poetry targets the neoliberal world of post-independence India by juxtaposing the cartographic global with the intensely local. But he goes farther. This essay shows that the cartographic impulse in Kolatkar's poetry is based on the poet's contradictory desire to achieve two concurrent yet opposite goals: one, to document the periphery of the modern world of Bombay/Mumbai (and therefore to make this subprime indigent life visible within authoritative contexts); two, to simultaneously also shield this periphery from the consuming eyes of the rest of the world (including from those of the reader of his own work). The poet's goal is to highlight the resistant edge and then make disappear this vehemently local element before it gets devoured by the exoticizing gaze of the global and the metropolitan.

> like a football player who while intent upon
> dribbling the ball may have a map of the
> entire field somewhere in his mind
> a precise map of the whole changing field
> at any given moment
> (Kolatkar, *Boatride* 237)

1

In many ways, the poet Arun Kolatkar (regarded as the foremost of his generation in both English and Marathi poetry in India) is not unlike the more famous postcolonial novelist, Amitav Ghosh. In one of his interviews, Ghosh remarked that the aim of all his writing was "to connect the resolutely panoptical and the irreducibly local" (Mahadevan-Dasgupta n. pag.) – the world in a grain of sand, but also the world *of* the grain of sand. Kolatkar's long poems function in a similar manner. Kolatkar connects the large to the small in a number of different ways – his books place smaller units of individual poems within a larger structure of a long poem; temporally, he juxtaposes myth and the pedestrian present; and, with regard to geography and location, he places the minutely local against larger cartographic structures. This essay examines the last of these, the cartographic impulse that pervades his bilingual poetry.

Arun Kolatkar's first book publication was *Jejuri*,[1] a volume in English that won the Commonwealth Poetry Award in 1977. This book charted the location of the various temples in this town of pilgrimage through the perambulations of the alienated poetic voice; it was a walking tour of this temple town. His next major book of poems in English, *Kala Ghoda Poems*, demarcates another spatial location in the city of Bombay, Kala Ghoda, and takes that as a starting point to explore the power differentials in the life of this metropolis. And, in between, he has many poems (in Marathi and in English) that name, locate and recreate travel routes, cities, towns, villages and monuments. This is a poet who is not just aware of, but obsessed with, the understanding of space; the process of mapping (in the literal and extended meaning of the word) provides the theme of the poems and also informs the structure of his books. The epistemology of cartography becomes the tool with which Kolatkar explores the modern post-independence Indian landscape in Bombay and its environs.

Maps are invariably connected with the large overview, the perspective of the establishment, and the voice of privilege. Notions of scientific facticity, impersonal knowledge-gathering and a final authoritative act of naming attach to the use of maps in the modern world. This can be seen right from the imperial documents that hierarchize the information gathered through different ways, with the mapmakers' knowledge being valued the highest:

> First […] were the reports of native travellers which shed a wide but uncertain light of the vast unknown. Behind them, piercing this gloom, came natural shoots of clear light representing the travels of individual European explorers. Finally, and well behind, came the zone of harsh white reality shed by the surveyors and map-makers. (Keay 188)

The descriptors used for map-makers as opposed to the "native travelers" suggest the valorizing of the surveyors' methods by depicting the map-makers as working in the unambiguous, unblemished white light of verifiable reality. Geographers such as J.B. Harley decry such unquestioning faith in the factual veracity of maps (both in the common perception but also in the study of geography itself) and point out that maps are authored texts and therefore as manipulable as any authored text. Harley demonstrates how the principles of cartography reflect "values, such as those of ethnicity, politics, religion or social class," and how "they are also embedded in the map-producing society at large" (156). He shows how the power of cartographic surveys emerges from a collocation of institutionalized/ imperial power, the authority of science and the force of the visual in the use of maps and reveals the "power of the map maker [that] was generally exercised not only over individuals but over the knowledge of the world made available to people in general" (166).

Others, such as Denis Wood and John Fels, show how all maps involved approximations and erasures that lead to a selective representation of the world, while Geoff King shows how such authored representations were put to use in western imperial projects. As de Certeau noted as early as 1984, maps remove the elements of the itinerary and everyday practice from their visual representation and adopt a totalizing, authoritative image in their end form; he pointed out "the bipolar distinction between 'map' and 'itinerary,' the procedures of delimitation or 'marking boundaries' […] and 'enunciative focalizations' (that is, the indication of the body within the discourse)" (116). The itinerary is directed by individual goals and follows a uniquely personal route whereas the abstraction and distanciation inherent in mapping strategies eliminates the individual or the local from the final document.

Since de Certeau and J.B. Harley, there have been many studies that have expanded the study of maps in other disciplinary areas and which took the Harley critique of cartography as the starting point. In postcolonial studies, Graham Huggan protests a one-sided reading of maps as tools of power or imperial design and shows how they can be expressions of counter-resistance as well. Sumathi Ramaswamy shows how maps are ideologically imprinted by analyzing how nationalist concerns were furthered by the merging of the national map and the image of Mother India during the anti-colonial movement and in postcolonial India. And Kapil Raj takes a variation on this theme when he shows how some cartographic ventures were co-authored by the locals and the colonial masters, thus muddying the one-way reading of victim and victimizer that is typically applied to colonized people; in this view, the mapping endeavor was a joint venture in many ways (Raj 13). This brief overview of theoretical studies suggests that maps are angular in their approach to the material world; they can be used to impose an outsider's world view on a native populace but also as a means of undermining the dominant ideology.

2

Arun Kolatkar represents maps on several levels in his work and highlights the ambivalent relation of maps and cartography to power structures. In both Kolatkar's Marathi and English poems, mapping becomes the structuring principle, the thematic core and a source of imagery. In fact, his "journey poems" became well known early in the poet's writing career (Mehrotra, Introduction, 19–20). The documentation of journeys, travels and movements of people and self is only one perspective on his work, which deals with the notion of hierarchy and power, and with ideas of freedom and imprisonment through the use of mapping structures.

There are different ways to explore Kolatkar's engagement with the process of mapping. One route would take us through the places the poet has mapped in his Marathi and English poetry, in order to see what pattern emerges from them. The reader can start with the mapping of the site of pilgrimage near Pune, the temple town of Jejuri; a gradual approach to the city and environs of Bombay[2] in "The Turnaround"; the interiors of the city, and the travel routes within them, in "The Barefoot Queen at the Crossroads"; the mapping of a particular "trisland"[3] in the district of Kala Ghoda in "Breakfast Time at Kala Ghoda". Alternatively, the reader can arrive at a similar understanding by examining the role of cartography in the various books of poems by Kolatkar: from the mapping of Bombay and its environs in *Jejuri* and *Arun Kolatkarchya Kavita* to the layout of Kala Ghoda in most of the poems in *Kala Ghoda Poems*. Either way, it is obvious that the poet is not just spatially and visually oriented but that one of the prime concerns of this poetry is locating, documenting and ironically unfixing binary notions of space.

Let us begin with *Kala Ghoda Poems*, since it makes the most sustained use of cartographic imagery. The name "Kala Ghoda" literally means "black horse" in Hindi and refers to the stone equestrian monument to Edward VII that was installed in this space by Sir Alfred Sassoon to commemorate the king's visit to Bombay in 1876. Today, the monument can be found in the City Zoo, Jijamata Udyan in Bombay. *Kala Ghoda Poems* underlines the fact that even today the space continues to be called by the name of the absent monument, indicating the continuity between the elitist policies of the colonial society and the postcolonial era.

Kala Ghoda Poems is made up of several poetic sequences, along with some freestanding poems. The first sequence of nine poems is mainly in the voice of and through the perspective of the pi-dog, the stray that occupies the Kala Ghoda trisland at dawn. The dog has replaced the stone horse that represents the colonial administration (rightfully so, the poet might add) and yet it also mocks this replacement process because this substitution is so incommensurate with the complex reality of post-independence Bombay. The colonial masters are and should be replaced by the commoner, represented by the dog, and yet the commoner is a stray dog, the animal that is slighted, kicked around and mocked by Indian society (*kutra* in Marathi and *kutta* in Hindi is a descriptor of insult in India). The appropriateness of the transfer from colonial to post-independence Indian commoner, the gap between the high status and visibility of the colonial versus the lowliness of the Indian common reality, the compassion for this downtrodden dog/commoner who has to traverse this gap in the present, and the amusement created because of the unbridgeability of this gap in reality as opposed to the grandiose dreams and the pronouncements of the dog – all of this is achieved at one fell swoop by placing the stray dog in the place of the now-missing stone horse of the British colonial administration.

This dog, with blotches of color on his body that make him a "pi-dog", carries the map of 17th-century Bombay on its back:

> I look a bit like
> the seventeenth-century map of Bombay
> with its seven islands
> […]
> – with a pirate's
> rather than a cartographer's regard
> for accuracy. (*Kala Ghoda* 16)

Kolatkar declares his structural principle in the book here – the book is not going to give the reader just a panoramic aerial view of the lives of the homeless in Kala Ghoda; rather, it will present a personalized and "tactical" account, as de Certeau defines the term, of the day in the life of an ordinary man through some close views into the lives of a few individuals. The pirate's map has gaps in its documentation, just like the visual gaps between the blotches on the dog's back. And it represents a selective and purposeful seizing of the tools of the establishment in order to reach one's own individual goals.

This distinction is similar to de Certeau's differentiation between "strategy" and "tactic" in his theory of everyday practices, where he studies how everyday acts of consumption – the "tactics" of walking, eating, etc. – resist the well-laid-out narratives of power that are structured from the top down. A tactic "make[s] use of the cracks that particular conjunctions open in the surveillance of the proprietary powers. It poaches in them. It creates surprises in them. It can be where it is least expected. It is a guileful ruse" (37).

In contrast, a strategy is impersonal and formal and imposes its grid on a city's life from above. It involves a mastery of time and space in achieving this "scientific" structure and hence is divorced from the material, day-to-day concerns of the inhabitants of the city. Such "strategic" planning of the city is invariably connected with the hierarchical, vertical gaze, with the panoramic vision exemplified in large cartographic projects. Subversion through everyday life consists in finding the gaps within these large structures, modes of travel and existence that cannot be visible at that distance – the remapping of the city through these daily routines of ordinary individuals. It makes visible what is hidden and, in that sense, delineates the cracks in the otherwise smooth and homogenous surface of the institutionalized narrative.

These alternative visualizations of the city are repeatedly attempted by Kolatkar in *Kala Ghoda Poems*. In poem 7 of the 31-poem sequence "Breakfast at Kala Ghoda", the poet seemingly gives us a plain list of the kinds of dishes that are served in the restaurants around Bombay:

They are serving khima pao at Olympia,
dal gosht at Baghdadi,
puri bhaji at Kailash Parbat,

aab gosht at Sarvi's,
kebabs with sprigs of mint at Gulshan-e-Iran,
nali nehari at Noor Mohammadi's,

baida ghotala at the Oriental,
paya soup at Benazir,
brun maska at Military Café,

upma at Swagat,
shira at Anand Vihar,
and fried eggs and bacon at Wayside Inn.

For yes, it's breakfast time at Kala Ghoda
as elsewhere
in and around Bombay.

– up and down
the whole longitude, in fact,
the 73rd, if I'm not mistaken. (87)

The list seems obvious and therefore inane: of course, each restaurant would serve a different variety of dishes, wouldn't it? But a detailed look at the dishes reveals the intricate and diverse patina of community in the heart of the city of Bombay because each dish declares the existence of a different social and communal group there. *Shira* and *upma* are typical of Marathi/Brahmin breakfast dishes while *baida ghotala* harks to the Parsi/Irani community, *aab gosht/nail nehari/paya* soup to the different Muslim communities and "fried eggs and bacon" to the Christian/anglicized groups. The multifarious nature of life in Bombay is emphasized, in this poem written at a time when the Shiv Sena, the radical regionalist political party, was in the ascendant and asking for a Marathi/Hindu-only Bombay city. Also, if one tries to locate the restaurants on a map of the city, they all fall along the main highway that cuts right through the center of the city. The list of dishes actually maps the diversity that exists side by side in the heart of Bombay and this is a different method of surveying the city from the abstract, top-down manner of the official map. Coming after the poems that list the food eaten in various parts of the world and of the country, this poem places the daily practices of the people of Bombay on the map, so to speak, but in a very idiosyncratic manner. De Certeau's ideas seem to come to life in this poem that gives one an understanding of the city, not from the distant, aerial view of the highways and the main thoroughfares but with an intimate view of the complex structure of social and cultural distinctions in this space.

Another poem that illustrates this informal and unconventional method of exploring the city is "The Barefoot Queen of the Crossroads", a sequence of four poems that presents to the reader a street-dwelling woman washing and dressing herself up in full sight of the rushing populace around the trisland. Poem 3 reiterates the step-by-step process of her draping a sari on her body (this is a semi-undressed woman in the street, whose body is available for the voyeuristic eyes of the passing world) and the stages of wearing the sari are mapped yet again across a straight line that cuts across the center of Bombay (the same trajectory used for the charting of the restaurants discussed in the earlier poem): Dadar, Parel, Lalbaug, Byculla, Bori Bunder, Flora Fountain and Kala Ghoda. The woman pleats the folds of the five-yard sheet of cloth that is the sari, in order to tuck it in at the waist, and each fold of the sari is one location on this line:

> she holds the sari away from her
> at arm's length
> at a halfway point along the border,
>
> from where it's a short walk
> to the belly
> for her three fingers and thumb,
>
> as they collect the sari
> along the way
> in neat accordion folds
>
> (flip flap, flip flap,
> Dadar, Parel, Lalbaug, Byculla, Bori Bunder,
> flip flap, Flora Fountain
>
> and flip, we come to Kala Ghoda,
> which is where
> we've been all along). (*Kala Ghoda* 78)

The pleating ends at "Kala Ghoda" on the map but it stops at her waist on her body, where the woman gathers the pleats together and folds them in just below her navel. The location in juxtaposition to the semi-undressed woman on whose lower waist the "black horse" (Kala Ghoda) is now mapped, acquires a surprisingly bawdy turn and this provides a completely surprising interpretation of the city and its people. The woman now seems to be holding the heart of the city of Bombay within her sari – all those named locations are right along the center of the city – and the poem presents the possibility that this woman is a prostitute. It also suggests the shocking prospect that all the seemingly "decent" middle-class people, who use these parts of the city for work and travel, her customers, might be complicit in the social and economic exchange. Earlier in the book, Kolatkar shows that the city of Bombay lives off the backs of the people who service it. For example, in the poems on the garbage sweeper Meera, Kolatkar writes how the piles of garbage that she collects by the roadside should each be rightfully seen as a homage to Bombay "since a good bit of city stands on sweepings such as these" (28). This refers to the geographical reality of Bombay, which stands partially on land redeemed from the sea. And suddenly, here, the city that stands on reclaimed land becomes also the city that stands on the labor of these poor, lower-caste sweepers. In this context, the woman wrapping the city in her sari now acquires new and unexpected meaning, with her sexual labor also being added to the equation. The poem criticizes the hypocrisy of the privileged who use and sexually abuse the street-dwelling workers, even as we see the tough woman declaring her non-victimhood in the way she purposefully wraps the city in her sari.

Such individualized cartography articulates a different vision of the space that is Bombay. According to the Census of 2001, the Greater Mumbai Municipality, along with the conurbations around it, has a total population over 28 million: "Over one half of these people, however, live in slums or are homeless; they live in shanty-towns. On pavements, along railway tracks, under bridges, and whatever other spaces are available to them" (Swaminathan 81). The city boasts of some of the highest-priced real estate in the world, as well as one of the biggest slums in the world, Dharavi. Many poets have written about this city of extreme contrasts – among them, Ezekiel, Dom Moraes, Jussawalla, Chitre, Patel, Daruwalla, Eunice de Souza, Manohar Oak, Narayan Surve, Namdeo Dhasal – and they remark on a poverty that lives cheek by jowl with objectionable shows of wealth and power. To quote Arjun Appadurai:

> The rich […] seek to gate as much of their lives as possible, traveling from guarded homes to darkened cars to air-conditioned offices, moving always in an envelope of privilege through the heat of public poverty and the dust of dispossession. (628)

The poor in the city are seen as dispensable yet necessary – they "dirty" the place and are seen as a threat to the hygienic as well as social life of the city. Therefore, so-called cleanliness drives, "beautification" drives, chase the poor out of the city spaces to the margins – witness those initiated in the 1990s by Sharad Pawar in Maharashtra or the ones carried out by Vilasrao Deshmukh in 2005 (Bunsha para 7). And yet the poor are also courted as potential voters and as cheap labor that holds up the infrastructure of the city.

Kolatkar exposes this hypocrisy in his poetry and creates an interstitial space in his work to reflect the reality of the city, where the workers and the poor have to make up/invent spaces to live in the forbidding urban sprawl – the space of the traffic island in *Kala Ghoda Poems* or the transitory spaces of rest found by his various traveling personas. Through such depictions of the place, Kolatkar proposes a history and a geography of the city that take into account the oppressions of the marginalized communities and also the domination and lack of concern of the privileged. These literary maps chart the differential in power relations in a place that refuses to completely succumb to such unjust separations.

Eating different kinds of food, wearing a sari, walking around – these are ordinary acts of consumption of the city life but cumulatively they expose the lacunae in the official view of the city. Walking, as de Certeau's work repeatedly states, is the everyday act of subversion, because it cannot be contained within the visual rationality of the official map: "A *migrational*, or metaphorical, city thus slips into the clear text of the planned and readable city" (93). Such a metaphorical city emerges from unexpected corners in the poems of Kolatkar, as can be seen from his many journey poems that travel within and around the city of Bombay.

There are numerous poems, in both English and Marathi, where the speaker walks through neighborhoods of Bombay and presents a reconstructed, refashioned space of the city that does not feature in the advertisement brochures or in the imaginary of the comfortable middle class. "The Turnaround" (*Boatride* 73–76) is one such poem.[4] On an enlarged map of the state of Maharashtra (of which Bombay is the capital), one could see that the poem follows the route from Bombay to Kopargaon via Kalyan, Nashik and Rotegaon and then back to Bombay. The actions of the traveler/speaker depicted in the poem (eating *vada pav*; defecating in the temple; selling the book of *abhang* (songs) by Sant Tukaram[5] for money; sleeping on the ground in the temple; losing footwear; asking strangers for drinking water) denote the life of a person who does not seem to be part of the race for middle-class prosperity, either because he is down on his luck or because he chooses to be outside that frame of reference:

> Bombay made me a beggar.
> Kalyan gave me a lump of jaggery to suck.
> In a small village that had a waterfall
> but no name
> my blanket found a buyer
> and I feasted on just plain ordinary water.

> I arrived in Nashik
> peepul leaves stuck between my teeth.
> There I sold my Tukaram
> to buy myself some bread and mince.
> When I turned off Agra Road,
> one of my sandals gave up the ghost. (*Boatride* 73)

Later, the speaker dispassionately lists his misfortunes – he has little to eat, he is dragged to court in another village, a dog dies near him in a roadside temple as he is lying down on the floor – but none of this elicits anguish or anger from him. The persona is alienated from every place he encounters as he travels out of Bombay and goes as far as Kopargaon before turning back. He does find a picture of pure and unadulterated simplicity and value in the form of a poor man and his daughter in Kopargaon; and yet, in the end, he chooses to "turn around" and return to Bombay.

> Water dripping down on my elbows
> I looked at the old man.
> The goodly beard.
> The contentment that showed in his eyes.
> The cut up can of kerosene
> that lay prostrate before him.

> Bread arrived unbidden,
> with an onion for companion.
> I ate it up.
> I picked up the haversack I was sitting on.
> I thought about it for a mile or two.
> But I knew already

> that it was time to turn around. (*Boatride* 76)

The walking journey documents the outside world of the city and its surrounding spaces and the way in which there seems to be a creeping commonality to the supposedly separate areas of the rural and the urban – the speaker appears to be equally disengaged from both the worlds in the poem. This sensibility can be explained by the familiar Marathi exclamation, used when one wants to express a difference of opinion from the statement just made but without any bitterness, with detachment, "Aso"; one could translate the word into English as "So be it." *Aso* is also the title of the little magazine that Ashok Shahane, close friend and publisher of all of Kolatkar's works, brought out in the early 1960s. This little magazine, like several others at the time, was a protest against the Marathi literary establishment that refused to heed the new writers' worldview and writing practices and the title indicated an abandonment of any hope or desire to be part of this outdated and unchanging world of publishers and writers. In his very influential essay on contemporary Marathi literature, *"aajkaalchyaa maraathi vaangmayaavar 'ksha' kiran"* (An X-ray of Contemporary Marathi Literature), Shahane explains why he named his little magazine *Aso* ("So be it"):

> If things go on as they have until now, there is nothing useful that will come out of it and at least for now we ourselves are not interested in giving any new direction to these events. Now there will be only one kind of a connecting handshake between them and us where each of us says "let it be" or "so be it". (29; my translation from Marathi).

So too with Kolatkar's speaker in the poem. From the outside, the small-town life surrounding Bombay might have seemed like an antidote to the alienating urban living in Bombay but it ends up being more of the same; the poetic voice has no use for either and therefore the speaker might as well say "aso" (so be it) to this estranging world.[6]

But the poem can be read two ways: as an unmitigated disengagement with the world (despite finding something of worth, the journey seems to change nothing); or, as suggested by Arvind Krishna Mehrotra, as a poem which finds some affirmation in the material substantiation of that world on the page. As Mehrotra writes, "*Mumbaine* [The Turnaround] […] is an internal journey as much as it is a real one, with real places (Kopargaon) and real events (Stalin's death)" (E-mail to author). There is a strong sense of the positive in the image of the old man in this poem which cannot fully be explained by the thematic of disengagement of the speaker from the world[7]. That excess of emotion can be explained as a marker of Kolatkar's metapoetic project here. Mehrotra points out how this poem, that seemingly shows the deadened sameness of rural and urban spaces, actually marks a point of departure in Kolatkar's oeuvre: the poet discovers a method of mooring his poetry in the material world in a way that avoids the indulgence of an exclusively internalized perspective (as in his earlier poems, that depended largely on images of surrealism) (E-mail to author). So while the speaker in the poem has not found anything of value that he can use in his own life, the poet, Kolatkar, has indeed discovered in this poem a method of anchoring his poems in the day-to-day material immediacy of the world. The inexplicable surplus of positive value that this poem embodies in the image of the old man (he stands out in glowing strangeness in the midst of images of the pedestrian and the ineffective) may be thus explained by taking recourse to the metapoetic aspects of the writing of this poem. The concrete evidence of the material world works as a counterweight to the speaker's withdrawal from it and it creates a poetry that simultaneously engages and disengages from the society which it describes. This documentary turnaround has been the distinctive feature of all of Kolatkar's later work.

This walking tour, then, brings to light a way of reading these spaces that is not part of the institutional charting of them – it makes visible the contradictions and the intricacies of the lives within. As de Certeau says, "Walking affirms, suspects, tries out, transgresses, respects, etc., the trajectories it 'speaks'. […] These enunciatory operations are of an unlimited diversity. They therefore cannot be reduced to their graphic trail" (99). What the city map shows as separate (the urban and the rural), this walking visit shows the two as both connected and disconnected in their deadened spaces.

The most ambitious of the poetic sequences in the book, and the one that attempts to show the multiple trajectories of social and economic relations that construct the space of the city, is "Breakfast time at Kala Ghoda". Here, Kolatkar shows different modes of knowledge, disparate ways of encountering the city, and finally the unknowable and therefore resistant nature of the space and the lives of the poor on the streets. The sequence comprises 31 poems, which initially provide an aerial, long-distance view of Kala Ghoda; it is interesting to see that, from this perspective, the people at Kala Ghoda are invisible:

> The clock displayed outside
> the Lund & Blockley shop across the road
> is the big daddy of all clocks,
>
> and will correct me if I'm wrong;
> but I think it's tonight already
> in Tokyo
>
> where they're busy polishing off
> sliced raw fish,
> sushi balls and tofu with soy sauce;
>
> and the emperor's chopsticks are poised,
> at this very moment,
> over Hatcho Miso, his favourite dish.
>
> In a restaurant in Seoul,
> a dog is being slowly strangled
> before it's thrown into a cooking pot. (*Kala Ghoda* 80)

Moving from the view of the entire world to the national view, the poem pans over Andhra Pradesh (poem 4) before settling on the larger view of Bombay (see poem 7 of the sequence, quoted earlier, on the food eaten by various communities in Bombay). And on its way, there is initially ample indication that this is a top-down, panoramic gaze at Bombay and at Kala Ghoda: for example, there are references (87) to the longitude on which Bombay is located ("the 73rd, if I'm not mistaken"). When it hovers on top of the world, the poem seems at first glance to be the "god trick" that Donna Haraway disavows in her essay on "situated knowledges": "I am arguing for the view from a body, always a complex, contradictory, structuring and structured body, versus the view from above, from nowhere, from simplicity. Only the god trick is forbidden" (588). However, Kolatkar demonstrates how the map can be "decolonized", to use Graham Huggan's terminology, by using this overview to locate the Kala Ghoda trisland in the middle of the world, to establish the relevance of this small place in the context of the USA, Japan, Korea, Russia and even the outer space; the poet insists that this trisland exists, just as those other better-known places do. The cartographic terminology ("the 73rd" longitude), the juxtaposition with better-known places in the world, all make the place, Kala Ghoda, and its inhabitants visible in the global setting. But very soon the poetic camera settles on a close-up of the people in the trisland – "the little vamp, the grandma, the blind man, / the ogress, / the rat-poison man, / the pinwheel boy, / the hipster queen of the crossroads [...] the pregnant queen of tarts" (*Kala Ghoda* 96) – who have all gathered around the idli-vendor[7] to get what would perhaps be their only meal of the day:

Each and every hungry and homeless soul
within a mile of the little island
is soon gravitating towards it

to receive the sacrament of idli,
to anoint palates
with sambar,

to celebrate anew, every morning,
the seduction and death
the demon of hunger (97)

The poems zoom in even further on the steaming idlis, the cheap food that seems like manna from heaven for these underprivileged people:

The tight lid
of the jumbo aluminium box
opens

with the collective
sigh
of a hundred idlis

waiting to exhale
followed
by a rush to the exit

– a landslide of fullmoons
slithering
past each other,

to tumble in a jumble,
and pile up
in a shallow basket,

an orgy,
a palpitating hill
of naked idlis

slipping and sliding
clambering over
and suffocating each other. (95)

The description of the sighing, eroticized idlis seems incongruous, and therefore humorous, because the food of the pavement dwellers is described in such grandiose terms and also because something so ordinary is given the extraordinary treatment; it also seems poignant, because these idlis are their only joy of the day; and beautiful, because it emphasizes the overall theme of the refusal to make these dispossessed people invisible or their bodily existence insignificant.

But before this turns into a voyeuristic spectacle where the street dwellers become the object of the reader's curious gaze and of his/her pity, the poet snatches away the brief glimpse into the private lives of these dispossessed people (that they are forced to lead in public view because of the lack of housing for them in the city). The idli-vendor closes up her mobile shop and moves on, the trisland inhabitants disperse, and our view of their life on the street has been as the joy of eating the idlis:

The pop-up cafeteria
disappears
like a castle in a children's book

– along with the king and the queen,
the courtiers,
the court jester and the banqueting hall,

the roast pheasants and the suckling pigs,
as soon as the witch
shuts the book on herself –

and the island returns
to its flat
boring self. (113)

This is the last poem in the sequence and, when it ends, it makes disappear this transient visual image of the place, thus making any totalizing visual representation of the people or their space unavailable for the reader's theoretical or visual pleasure. They remain unknowable in the end and beyond the grasp of homogenizing and universalizing structures of representation. The book of images folds on itself and the reader/viewer is left with nothing.

The impulse behind this "letting go" is similar to what the modernist poet Marianne Moore describes in her poem on the jellyfish:

Visible, invisible,
a fluctuating charm
an amber-tinctured amethyst
inhabits it, your arm
approaches and it opens
and it closes; you had meant
to catch it and it quivers;
you abandon your intent (Moore 66)

The poet wants to hold the delicate jellyfish in her grasp, but it quivers (showing its individual existence in that typical movement, and also reminding the poet of its sting) and she gives up the desire to hold it in her hand. "Visible/invisible", then, is the description of this delicate, elusive reality that swims away from one's eyes in an instant, just like the view of the community of the homeless in Kala Ghoda; it is equally futile, Kolatkar seems to imply, to hold them within the clamps of one's vision. It is also like the butterfly in the poem from *Jejuri*, where the words are incapable of encompassing the evanescent reality of the insect's fluttering, momentary life.[8] And this is the reason de Certeau says that writing/viewing petrifies its object.[9] The jellyfish swims away; the butterfly vanishes from the page; the inhabitants of Kala Ghoda disperse; the fairy-tale book folds up, and the reader's grasp is left empty – these people's lives are not available for entrapment.

Kolatkar's poetry thus valorizes the everyday acts of living of the ordinary, non-privileged people in the city of Bombay by undermining, in different ways, the official view of the state that disregards the lives of the poor. He presents alternative ways of mapping this space (through walking, eating, clothing oneself), by presenting the view from below, literally (as in the case of the pi-dog of Kala Ghoda) and by giving the reader oscillating perspectives on spaces that destabilize fixed notions of place and identity, like that of the football player who has "a precise map of the whole changing field / at any given moment" (*Boatride* 237). The map is transformed from a static, overarching and unchanging view to a dynamic, fluctuating understanding that attempts to represent the ever-changing surrounding world. Kolatkar does not abandon the map or its tropes – his poetry in both English and Marathi is replete with these images. What he does is to provincialize the practice of cartography and take away the claims of universalism from such mapping projects. He creates room for maps from other perspectives, from an angle to the established narrative. In the end, he mocks the authoritarian view of the colonizing map, of the cartographic desire of power, even as he documents the resistance of the non-privileged.

Kolatkar has been accused by nativist critics such as Rege and Kimbahune of unrooted cosmopolitanism – in effect, of taking a panoramic view of Bombay and its people. He has also been accused of exoticizing the Jejuri landscape in order to appeal to the "foreign" tastes of his English readers (Nemade 126) but an examination of this cartographic impulse in his writing shows that, as in "Breakfast Time at Kala Ghoda", the poet brings the entire world to bear upon his own particular location in Bombay. This is a poetic space that does not partake of the essentialist corner of nativist exclusion (like that advocated by Nemade) nor of the condescension of institutionalized and anglicized ignorance of regional diversities. The poet is like the pi-dog, with heritages in both a Sanskritized nativist elitism and an impossibly foreign colonial exclusionary lifestyle, but who makes his individual and (literally) pedestrian world out of these two extremes. Here is an alternative to both, a poetry that is able to represent, if only briefly, the subaltern that resides confidently on the border and in the street. In 1997, Sumit Sarkar criticized the "binary of Westernized surrender/indigenist resistance" (106) that marked the *Subaltern Studies* in India and asked for a more complex understanding of the native and the foreign; Kolatkar's poetry can be seen as one example of how such a binary is exploded, through the co-constitutive nature of the local and the global in his poetry.

The focus in Kolatkar's work is always on the local and the particular, even as it interacts with the global. In the unpublished manuscripts of the *Kala Ghoda Poems*, there are multiple drafts of the published version of the poems, some written in Marathi as well as English (as was the poet's wont – he frequently wrote in both languages). One of them depicts the young sanitation worker, Meera, for whom the *Kala Ghoda Poems* indicate a future in prostitution (since that is where most such kids end up in reality). But for now, in the poem, this girl could not care less about the world revolving around her. She is watching her mother striking a deal with the cop across the street, and Meera has got the street dirt on her bare behind as she plays on the roadside. The poet says, addressing her in utter admiration:

tujhyaa dhunganaana nuktaach prasiddha kelaa aahe

dhuliina chhaaplelaa jagaachaa nakaashaa (Kolatkar, unpublished manuscript n. pag.)

(Your buttocks have just published/publicized
the map of the world printed by the dust) (my translation)

The map of the world is on the little girl's bare buttocks – the irreverence and the disregard shown by the poet for this massive cartographic venture could not be more apparent. It mocks the power of the universalist cartographic projects by highlighting the incongruence of the provenance of the world map and its current home on the girl's behind. Perhaps this is the best example of how the resolutely panoptical gaze of the privileged cartographer meets the irreducibly local in the particularized behind of the homeless girl: the poet might as well exclaim, "The world, my ass!" And even as the two are inextricably connected, the local "moons" the global in this poetic cartography.

Acknowledgements

I am grateful to Mrs Soonoo Kolatkar for giving access to the Kolatkar archive and for permission to quote from and translate parts of Kolatkar's published and unpublished writing, and to Ashok Shahane and Michael Perreira for making this access possible. I must also thank the anonymous *Journal of Postcolonial Writing* reviewers of my article for several valuable suggestions.

Notes

1. Six years after his death in 2004, Kolatkar's manuscript for the Marathi book of poems on the same place/theme was published by Pras publications. All references to *Jejuri* here are to the English edition, unless otherwise specified.
2. The city of "Bombay" was officially renamed "Mumbai" (its Marathi name) after an agitation by Hindu nationalists in the 1990s. Kolatkar continued to refer to the city as "Mumbai" in Marathi and "Bombay" in English (he did not accede to the larger parochial demands of these regional political parties and this might have been one way of indicating that opposition). To indicate that practice, I will also continue to do the same in this essay, even though "Mumbai" is now the broadly accepted name in both English and Marathi.
3. The "trisland" is a word used by Kolatkar to describe the triangular space of the traffic island at Kala Ghoda.
4. For a detailed account of the origin of the poem and its relation to the poet's biography, see the Arvind Krishna Mehrotra's introduction to *The Boatride and Other Poems* (14–19).
5. Sant Tukaram (1609-50) is a popular saint in the state of Maharashtra and his *abhangs* (songs) are part of the everyday culture in the region. See Chitre and Kolatkar (*Boatride*) for an English translation of his songs.
6. In a way, this is connected to the theme of Kolatkar's first book of poems, *Jejuri*, which also depicts the walking journey of a poetic persona through the temple town of Jejuri, perhaps in search of some meaning not available in the urban life of Mumbai; the poet finds that (other than the butterfly and the excited fowl in the field) this site of pilgrimage is as empty of significance as the consumerist world of Bombay. See Amit Chaudhuri's introduction to the NYRB edition of *Jejuri*.
7. Idlis are steamed cakes of black lentil and rice that are eaten as a snack, or breakfast. They are usually paired with (and dunked in) sambar, a spicy lentil soup.
8. See Nerlekar for a detailed reading of this poem.
9. "The time is thus over in which the 'real' appeared to come into the text to be manufactured and exported. [...] It is no more than an illusory sacrament of the real, a space of laughter at the expense of yesterday's axioms" (de Certeau 152).

Works cited

Appadurai, Arjun. "Spectral Housing and Urban Cleansing: Notes on Millennial Mumbai." *Public Culture* 12.3 (2000): 627–51.

Bunsha, Dionne. "A Tale of Two Mumbais." *Frontline* 22.7 (12–15 Mar. 2005). Web. 30 May 2006.

Chaudhuri, Amit. Introduction. *Jejuri*. By Arun Kolatkar. New York: NYRB, 2005.

Chitre, Dilip. *Says Tuka: Selected Poetry of Tukaram*. New Delhi: Penguin, 1991.

de Certeau, Michel. *The Practice of Everyday Life*. Berkeley: U of California P, 1984.

Haraway, Donna. "Situated Knowledges: The Science Question in Feminism and the Privilege of Partial Perspective." *Feminist Studies* 14.3 (1988): 575–99.

Harley, J.B. "Deconstructing the Map." *The New Nature of Maps: Essays in the History of Cartography*. Ed. Paul Laxton. Baltimore: Johns Hopkins UP, 2001. 150–68.

Huggan, Graham. "Decolonizing the Map." *The Post-Colonial Studies Reader*. Ed. Bill Ashcroft, Gareth Griffiths, and Helen Tiffin. London: Routledge, 1995. 407–11.

Keay, John. *Explorers of the Western Himalayas: 1828–1895*. London: John Murray, 1996.

Kimbahune, R.S. "From Jejuri to Arun Kolatkar." *New, Quest* (Jan.–Feb. 1980): 27–32.

King, Geoff. *Mapping Reality: An Exploration of Cultural Cartographies*. New York: St Martin's P, 1996.

Kolatkar, Arun. Archive of unpublished manuscripts in Mumbai. Private collection.

———. *Arun Kolatkarchya Kavita*. Mumbai: Pras Prakashan, 1977.

———. *The Boatride & Other Poems*. Mumbai: Pras Prakashan, 2009.

———. *Jejuri*. Mumbai: Pras Prakashan, 1976.

———. *Jejuri*. New York: NYRB, 2005.

———. *Kala Ghoda Poems*. Mumbai: Pras Prakashan, 2004.

Mahadevan-Dasgupta, Uma. "Translating the World." *Hindu*. 29 Aug. 2004. Web. 8 June 2005.

Mehrotra, Arvind Krishna. Introduction. *Boatride & Other Poems*. By Kolatkar.

———. E-mail to the author. 22 July 2011.

Moore, Marianne. *A Marianne Moore Reader*. New York: Viking, 1961.

Nemade, Bhalchandra. "Arun Kolatkar: Bilingual Poet." *Tikasvayamvara*. Aurangabad: Saket Prakashan, 1990. 117–26.

Nerlekar, Anjali. "The Rough Ground of Translation and Bilingual Writing in Arun Kolatkar's *Jejuri*." *Perspectives: Studies in Translatology*. 12 Mar. 2012. Web. 24 Apr. 2012. <http://www.tandfonline.com/doi/abs/10.1080/0907676X.2011.649771>.

Raj, Kapil. *Relocating Modern Science. Circulation and the Construction of Knowledge in South Asia and Europe, 1650–1900*. Basingstoke: Palgrave Macmillan, 2007.

Ramaswamy, Sumathi. *The Goddess and the Nation: Mapping Mother India*. Durham, NC: Duke University Press, 2010.

Rege, Pu. Shi. "Jejuri." *Rucha* 1–2 (Kolatkar Spec. issue, 12 Sept. 1977): 76–78.

Sarkar, Sumit. "The Decline of the Subaltern in *Subaltern Studies*." *Writing Social History*. New Delhi: OUP, 1997. 82–108.

Shahane, Ashok. "An X-ray of Contemporary Marathi Literature." *Napeksha*. Mumbai: Lokvangmaya Griha, 2008. 3–30.

Swaminathan, Madhura. "Aspects of Poverty and Living Standards." *Bombay and Mumbai: The City in Transition*. Ed. Sujata Patel and Jim Masselos. New Delhi: OUP, 2003. 81–109.

Wood, Denis, and John Fels. *The Power of Maps*. New York: Guilford Press, 1992.

Index

INDEX

ummah, 56
Urdu, 98

Vogel, Pierre, 52

Waldman, Myron, 66
Waymo, 52, 53
Web 2.0, 52; Islam, relationship between, 55–6
Western pop culture, 50, 74
women, stereotypes of, 84–5
World Economic Forum, 4
World Have Your Say (WHYS), 18, 22–4, 25–6, 30, 31
World Laughter Day, 74
World Social Forum, 2; birthplace of, 7; communication technologies, use of, 8, 9, 13, 14; communication, politics of, 8–9; criticisms of, 5; decentralization of, 6–7; dynamics of, complex, 10; globalizing of, 5–6; grassrooting of, 6; networking, 8; open space of, 5; overview, 4–5; place, significance of, 9, 12–13; scale, significance of, 13; transformation, social, 11–12; *see also* Global Day of Action (GDA)

Yahya, Mohammed, 54
Yakusho, Koji, 82, 89
YouTube, 52

Zentralrat der Muslime (Central Council of Muslims), 52